Oscar Codling 11/3

Claire Hutchinson and Danny Foulder

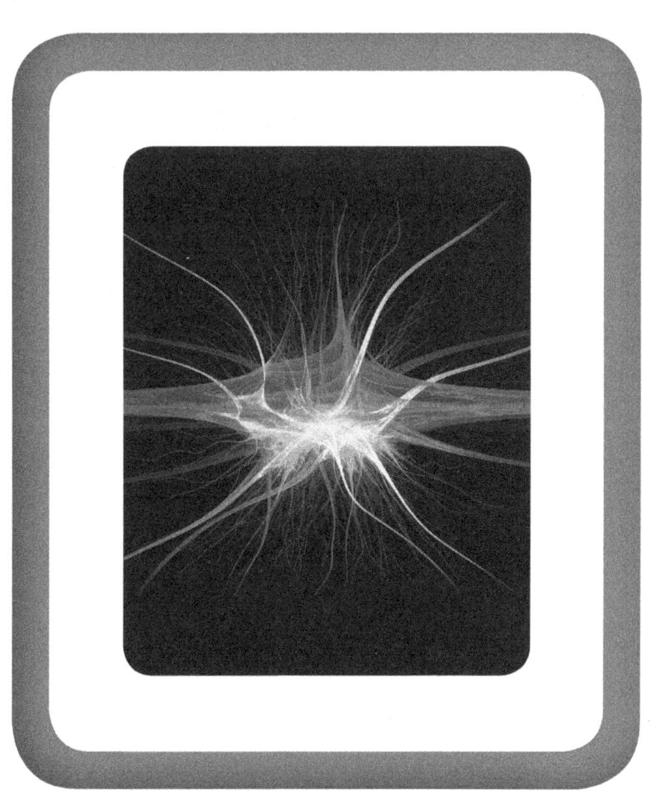

ESSENTIALS

OCR GCSE Physics B

Contents

1. The unit used to measure temperature is degrees Celsius (°C). What is the unit of **energy**? [1]

2. This image uses colour to represent temperature.

(a) Name this type of image. [1]

Thermal image

(b) From where is the house losing the most heat energy? [1]

~~win do~~ windows (1)

3. Look at the table of data.

Material	Specific heat capacity (J/kg°C)
Copper	380
Aluminium	880
Water	4200
Titanium	523
Silver	233
Helium	5193

(a) Which material will require the most energy to raise its temperature by 1°C? Explain your answer. [2]

~~Silver~~ Helium (1)

(b) Calculate the energy needed to raise the temperature of 2 kg of copper by 10°C. [2]

7600 (1)

4. Julie heated some chocolate. She measured the temperature regularly and plotted her results on the graph below.

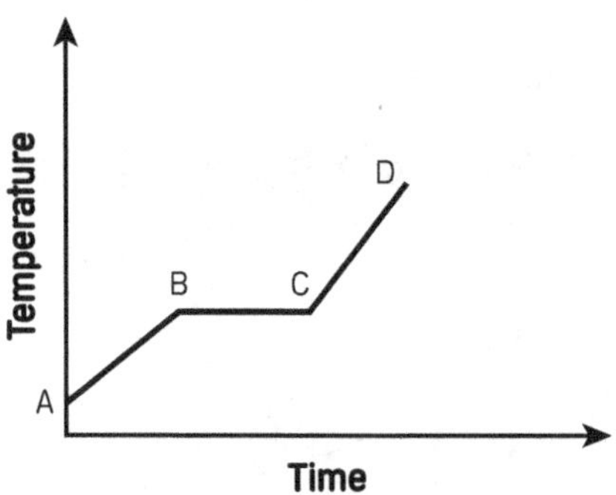

Explain what is happening to the chocolate between points **B** and **C**. [2]

It starts Cooling Down

5. The amount of energy needed to melt or boil a material is called the **specific latent heat**.

(a) What **two** factors does a material's specific latent heat depend on? [2]

Mass of the objects **and** temprature

(b) The specific latent heat of ice is 330 J/g. How much energy is required to melt 20 g of ice? [2]

6600

[Total: .... / 13]

Higher Tier

6. The specific heat capacity of iron is 440 J/kg°C. On a sunny day a section of railway track heats up from 10°C to 25°C. This requires 660 kJ of energy. What is the mass of the railway track? [2]

7. Distinguish between temperature and heat. [2]

8. Harry carries out an experiment using 0.5 kg of paraffin wax. It takes 7500 J of energy to raise the temperature by 6°C. Calculate the specific heat capacity of paraffin wax. [2]

150,000

9. Sadiq heats 200 g of ice. The graph shows his data.

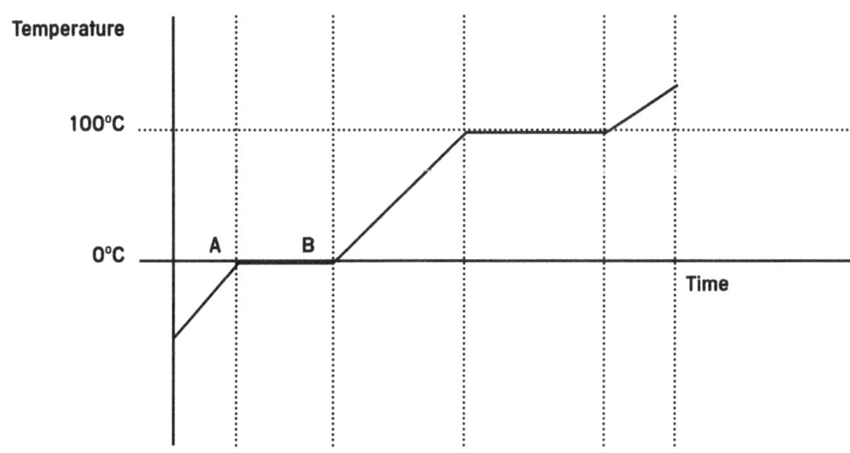

(a) Energy is supplied between points **A** and **B**. Explain why the temperature of the water does not change. [2]

(b) The energy required to melt this ice is 66.8 kJ. Use this information to calculate the specific latent heat capacity of melting water. [2]

[Total: .... / 10]

P1 Keeping Homes Warm

1. Describe what happens to infrared radiation when it comes into contact with

(a) a shiny surface. .... [1]

(b) a dull or rough surface. .... [1]

2. The table below shows the initial cost and annual saving on energy bills of three different energy-saving methods.

Method	Initial cost £	Annual saving on energy bills £
Double-glazing	3000	200
Loft insulation	200	400
Cavity wall insulation	1000	500

(a) Which of the above methods reduces heat loss by the most in a year? [1]

....

(b) Explain which of the above methods has the shortest payback time. [2]

....

....

3. Jane uses a hairdryer. Some of the energy is wasted as sound. The sankey diagram shows the total energy input and the output energies.

Electrical energy input 30 000 J

Thermal energy .... J

12 000 J wasted

(a) Calculate the thermal energy output. [1]

....

....

(b) Calculate the efficiency of the hairdryer. [2]

....

....

(c) A different hairdryer has an efficiency of 50%. Is this more or less efficient than Jane's hairdryer? [1]

less

4. **(a)** Light bulb A uses 55 J of energy per second and gives out 11 J of light energy. Light bulb B uses 50 J of energy per second and gives out 5 J of light energy. Light bulb C uses 60 J of energy per second and gives out 18 J of light energy. Which light bulb is the most efficient? [3]

A = 55 − 11 = 44J wasted B = 50 − 5 = 45J wasted C = 60 − 18 = 42J wasted so C is the most efficient

(11 ÷ 55) x 100 = 20% (5 ÷ 50) x 100 = 10% (18 ÷ 60) x 100 = 30%

B = 10% efficient A = 20% C = 30% efficient

(b) What is the difference in efficiency between the most and least efficient light bulbs? [1]

Differnce of 20% efficency

5. Explain how the design features of a home can help to prevent heat loss. [6]

The quality of your written communication will be assessed in this question.

[Total: ____ / 19]

Higher Tier

6. Name and describe the **three** ways energy is transferred. [6]

7. A light bulb has an efficiency of 80% and a total energy input of 80 J per second. How much energy does the light bulb waste in an hour? [3]

[Total: ______ / 9]

1. Look at the diagram of a wave.

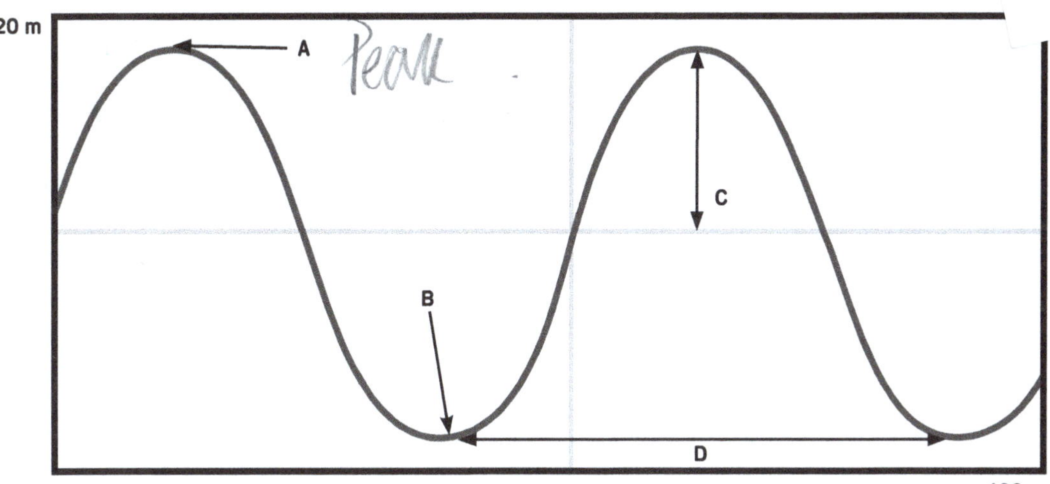

(a) Write down the correct name for each of the features labelled on the diagram of a wave. [4]

A Crest **B** trough **C** Amplitude **D** wavelength

(b) James counts the number of waves which pass a point in 30 seconds. James counts 120 waves. Calculate the frequency of the wave. [2]

120 ÷ 30

4 waves per Second

(c) Calculate the wave speed of the wave shown above. [2]

2. Fill in the blanks below to show the types of electromagnetic waves that make up the spectrum. [5]

radio waves ....

.... **gamma**

[Total: ...... / 13]

Higher Tier

3. The speed of sound in air is measured to be 330 m/s. Calculate the wavelength of a sound wave which has a frequency of 194 Hz. [2]

[Total: ...... / 2]

Light and Lasers

1. Morse code can be used to send signals. How is light used to send a message using Morse code? [2]

....

....

2. The diagrams below show what can happen to a ray of light when it is incident on a glass-air boundary.

A

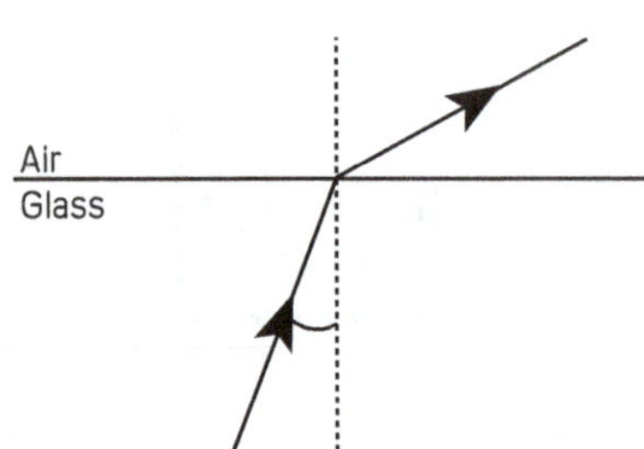

B

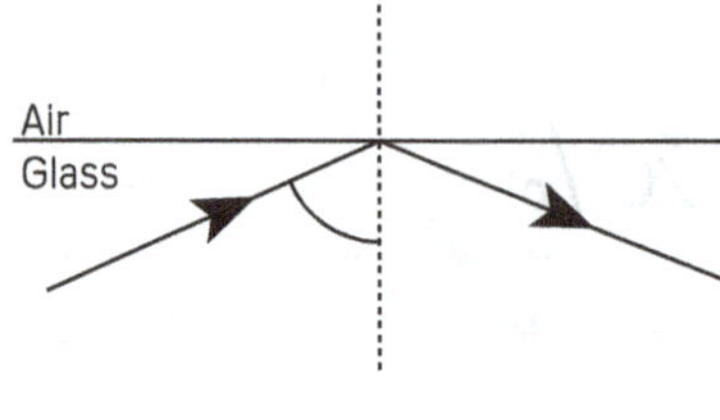

C

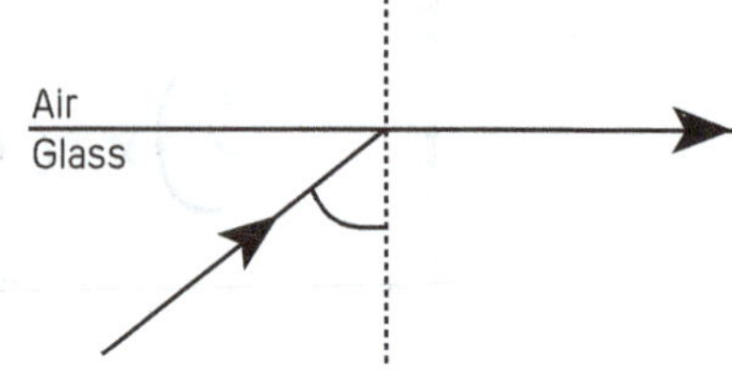

(a) Which picture, **A, B** or **C**, shows what will happen to the light when,

(i) the angle of incidence is **greater than** the critical angle? [1]

....

(ii) the angle of incidence is **equal to** the critical angle? [1]

....

[Total: .... / 4]

Higher Tier

3. Describe how data is stored on a compact disc and how this data is read by a laser. [4]

....

....

....

....

[Total: .... / 4]

1. Microwaves are part of the electromagnetic spectrum. Microwaves can be used to cook food.

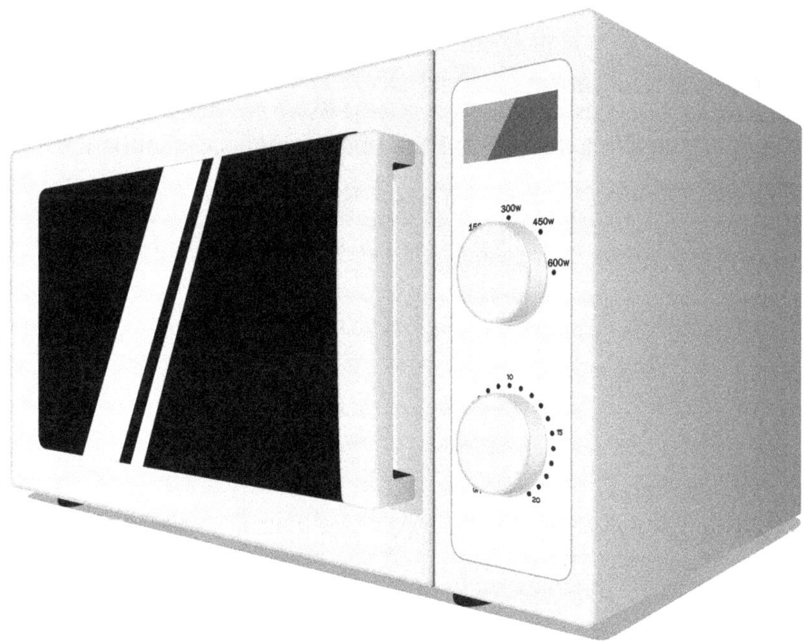

(a) Which **two** substances in food absorb the energy from microwaves? [2]

protein WATER **and** FAT

(b) Approximately how far do microwaves penetrate into the food? [1]

into the Centre 1cm

(c) Which other part of the electromagnetic spectrum is often used to cook food? [1]

infa red ✓

2. **(a)** A large mobile phone mast is being built near a group of houses. Some of the residents are worried about this while others don't think it's a problem.
Explain why the residents have these two different viewpoints. [3]

(b) What evidence could the residents use to help them form an opinion on the safety of the phone mast? [2]

[Total: ...... / 9]

Higher Tier

3. Microwaves can be used to transmit signals.

(a) Some houses are unable to receive mobile phone signals. Describe the conditions that will affect a microwave signal. [4]

(b) Describe **two** ways in which these problems can be reduced. [2]

4. A common misconception is that microwave ovens cook food 'from the inside out.' Explain why this is not the case. [5]

[Total: ...... / 11]

1. Infrared is a type of electromagnetic radiation. Write down **two** uses of infrared radiation. [2]

lights **and** Cooking

2. These pictures show signals.

A

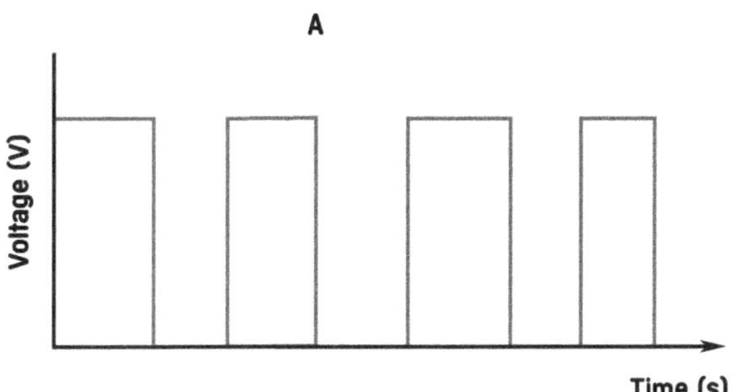

B

Voltage (V)

Time (s)

A Analog B Dialog

(a) (i) Label the signals. [1]

(ii) How did you come to this conclusion? [2]

(b) Which of the above signals could noise be most effectively removed from and why? [2]

A ~~its~~ because its only 1s and 2s

(c) The diagram shows a signal. Write the code for this signal in the boxes below. [1]

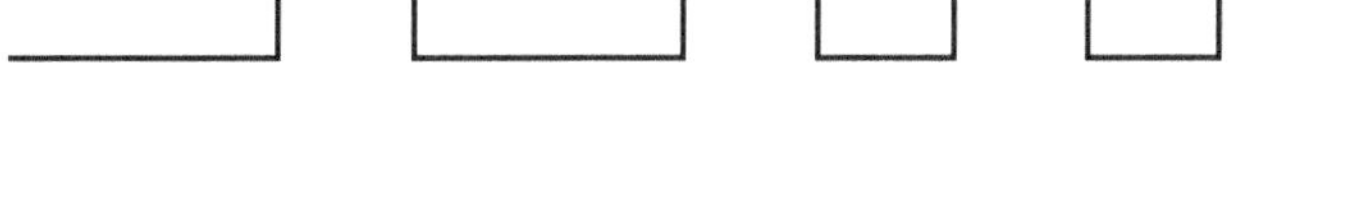

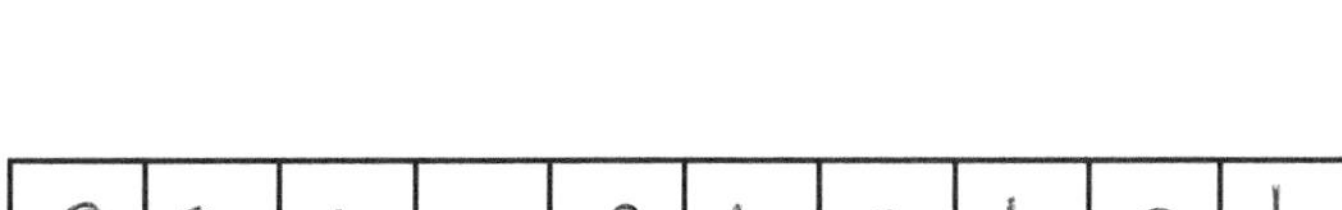

0	0	1	0	0	1	0	1	0	1

1

[Total: ...... / 8]

Higher Tier

3. Multiplexing can be used to send digital signals.

(a) What is multiplexing? [1]

(b) State **two** benefits of multiplexing. [2]

[Total: ...... / 3]

1. Many modern devices use wireless technology.

(a) Write down **two** uses of wireless technology. [2]

~~its portable~~ portable radio

~~easy to use~~ Console Controllers

(b) Write down **two** benefits of wireless technology. [2]

its portable

~~easy~~ More Convenient

2. Simon is writing about DAB radio.

Put a tick (✓) next to each of the statements that are correct and a cross (✗) next to each of the statements that are wrong. [4]

Anyone with a DAB radio can receive digital radio. ✓

DAB radios can receive digital broadcasts and old analogue signals. ✓

There are far more digital stations available than analogue stations. ✓

Digital signals sometimes contain more interference than analogue signals. ✗

3. Listeners of a radio station have been complaining of poor reception. Another local station, F-FM, transmits on a very similar frequency.

(a) Explain the cause of the poor reception. [2]

another station emits on an similar frequency Causing interference with the two stations

(b) Give a possible solution. [1]

~~get it x~~ get a dab radio

[Total: .... / 11]

Higher Tier

4. Radio signals from France are often heard on the South coast of England. Describe how the ionosphere enables these signals to be transmitted around the curve of the Earth. [2]

....

....

....

5. Look at the house in the picture. This house is able to receive analogue radio broadcasts.

(a) Explain why it cannot receive shorter wave signals such as digital radio or mobile phone signals. [3]

....

....

....

(b) Suggest why the analogue signal will not be as clear as an analogue signal received on top of the hill. [2]

....

....

[Total: .... / 7]

1. This question is about seismic waves.

(a) Name the device that detects seismic waves. [1]

....

(b) P-waves are seismic waves that travel through the Earth at speeds of 5–8 m/s. What are the main differences between P-waves and S-waves? [3]

....

....

....

....

2. The Sun produces ultraviolet waves. Some ultraviolet waves can pass through the atmosphere.

(a) How is ultraviolet harmful to humans? [1]

....

(b) Ben has sun block with an SPF of 20. What is meant by **SPF**? [1]

....

(c) For how long can Ben stay out safely in the Sun when wearing his sun block?

Complete the table below. [3]

Safe time in the Sun	
Without sun block	With sun block (SPF 20)
3 minutes	60 minutes
10 minutes	200 minutes
25 minutes	500 minutes

3. Over many years scientists measured the levels of ozone gases in the upper atmosphere. From these measurements they concluded that a hole was developing. In what **two** ways could the scientists ensure their results were reliable? [2]

....

....

[Total: .... / 11]

Higher Tier

4. The diagram shows P-waves and S-waves through the Earth.

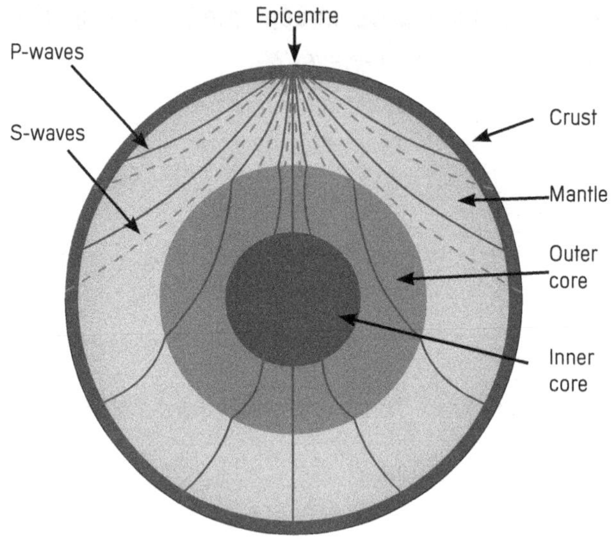

S-waves cannot be detected on the opposite side of the Earth to the epicentre of an earthquake. Describe why S-waves cannot be detected and explain what this tells us about the structure of the Earth. [4]

....

....

....

....

5. Explain the effect that the discovery of the hole in the ozone layer over Antarctica has had on society at an international level. [5]

....

....

....

....

....

[Total: ..../ 9]

1. This question is about photocells.

(a) What type of energy do photocells absorb? [1]

(b) Photocells can be joined together to form a solar panel. Describe the benefits of doing this. [3]

(c) Give **three other** ways in which the Sun's energy can be harnessed. [3]

2. **(a)** Sarah is a homeowner. What are the **advantages** of attaching electricity-generating photocells to the roof of her house? [3]

(b) What factors should she consider before having photocells fitted? [2]

3. The UK could generate all of its electricity needs from offshore wind power. What would be the main **advantages** and **disadvantages** of this? [5]

[Total: .... / 17]

Higher Tier

4. The diagrams show an experiment with a small solar cell.

A

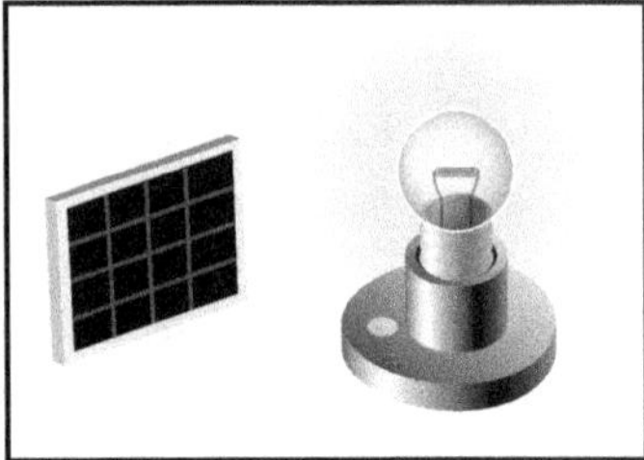

B

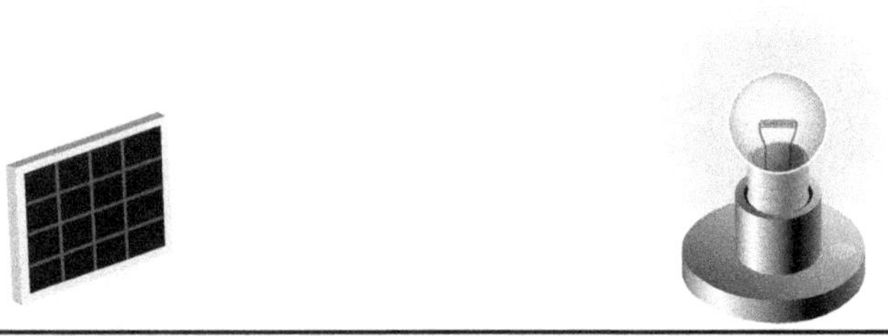

(a) Explain why the output of the solar cells would differ in positions **A** and **B**. [2]

(b) What else could affect the output? [1]

(c) Passive solar heating and a photocell both use the Sun as an energy source. Describe how these two processes differ. [6]

[Total: .... / 9]

1. **(a)** Describe how to generate electricity using the dynamo effect. [2]

....

....

(b) How could the effect be increased? [2]

....

....

2. Look at the graph below.

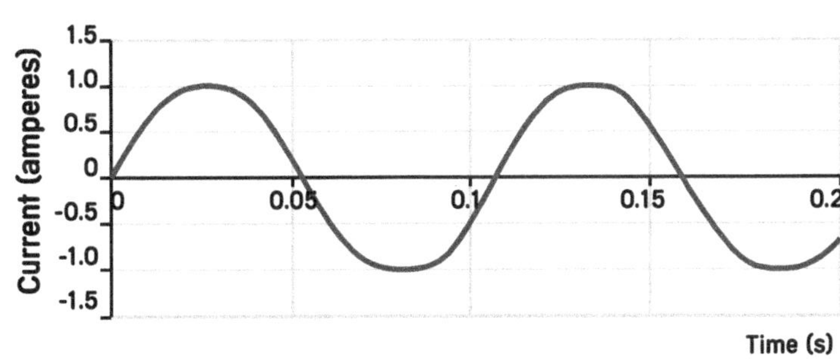

(a) What kind of current is shown in the graph? Explain your answer. [2]

....

....

(b) What devices generate this type of current? .... [1]

3. Electricity can be produced using fossil fuels. Describe the main stages by which electricity is produced in a conventional power station. [2]

....

....

....

4. When electricity is produced in a power station some of the energy is wasted as heat. A power station uses 12 MW of energy from coal. Only 3 MW of electrical power is produced. Calculate the efficiency of the power station. [2]

....

....

[Total: .... / 11]

Higher Tier

5. The picture shows a power station. As coal is burned in the power station energy is released. The power station wastes 375,000 J of energy as heat in the boiler, turbine, generator and cooling towers.

Overall this power station is 0.25 (25%) efficient. Calculate the total input energy released from the burning coal. [3]

....

....

....

....

[Total: .... / 3]

1. The average surface temperature of the Earth has risen rapidly over the last 100 years.

Over the last 100 years the levels of solar radiation reaching the Earth have increased. Carbon dioxide concentration in the atmosphere is increasing and one of the factors that is causing this is thought to be increased burning of fossil fuels. Increased carbon dioxide concentration in the atmosphere leads to the greenhouse effect.

Using the above evidence give an argument that supports man-made global warming and one which refutes it. [4]

2. What does **deforestation** mean and why does it lead to global warming? [3]

3. Give **two** examples of greenhouse gases. [1]

[Total: .... / 8]

Higher Tier

4. Explain how it is possible to have good agreement between scientists about the greenhouse effect but disagreement about whether human activity is affecting global warming. [6]

✎ *The quality of your written communication will be assessed in this question.*

5. This graph shows the mean global temperature from 1880 to 2000.

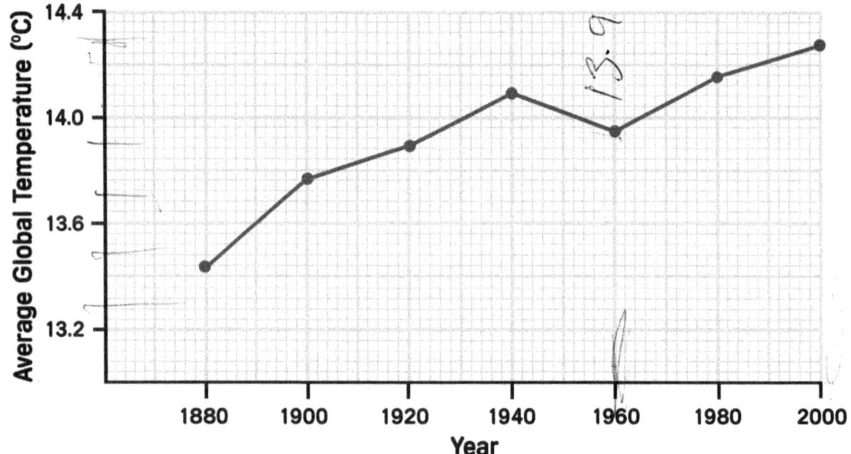

(a) Describe the pattern of global warming shown by the graph. [3]

(b) By how many degrees Celsius did the temperature rise between 1960 and 2000? [1]

[Total: / 10]

1. Complete the following table. [6]

Type of fuel used in power stations	Example of fuel	Advantage	Disadvantage
Fossil fuels			
Renewable biomass			
Nuclear fuels			

2. A hairdryer has a current of 10 A through it when plugged into the mains supply (230 V). Calculate the power of the hairdryer. [2]

3. Electrical appliances use a lot of energy. The energy used by devices such as hairdryers is measured in kWh.

(a) What does the number of kWh of energy used by a device depend on? [1]

(b) The lamp shown below uses 4 kWh of energy. Each unit of energy costs 10 pence. Calculate the cost of using this lamp. [2]

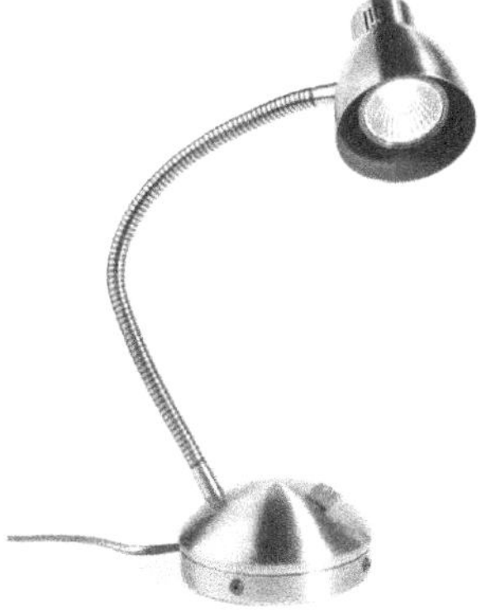

[Total: ...... / 11]

Higher Tier

4. **(a)** Complete the table below. [4]

Appliance	Power rating in kilowatts	Average time used each week (hours)	Total energy used in a week (kWh)
Hairdryer	2.3	1	
Dishwasher	1.2		3
Washing machine		4	6.8
TV	0.12		1.2

(b) The hairdryer is connected to the 230 V mains. Calculate the current passing through the appliance when it is switched on. [2]

5. Off-peak electricity, like Economy 7, is becoming popular again. Write about the **advantages** and **disadvantages** of off-peak electricity. [4]

Advantages

Disadvantages

6. When a large current is passed through a wire, energy is lost as heat. What are the implications of this for the transmission of power in the national grid? [4]

[Total: / 14]

1. Lucy is a radiologist in a hospital. As part of her job she must use radioactive chemicals.

   What precautions should Lucy take when handling radioactive substances? [4]

2. Suggest **one** beneficial use for each of the types of radiation listed below. [3]

   Alpha

   Beta

   Gamma

3. This question is about absorption.

   (a) What material can be used to stop beta? [1]

   (b) Which type of radiation is stopped by a few centimetres of air? [1]

   (c) Which type of radiation is able to pass through concrete? [1]

4. Which type of radiation is the least ionising? [1]

5. When does nuclear radiation form… [2]

   Positive ions

   Negative ions

6. Spent fuel is removed from a nuclear reactor core.

   (a) What element is removed from spent fuel? [1]

(b) Describe how low- and high-level waste is disposed of. [2]

Low-level waste ....

....

High-level waste ....

....

[Total: .... / 16]

Higher Tier

7. The table below shows the results of an experiment investigating the penetration of radiation from three different sources through a selection of different materials (× = radiation detected).

Source	One sheet of paper	Several pieces of thick cardboard	Thin sheet of aluminium	Thick piece of lead
1	×	×		
2	×	×	×	
3				

(a) Identify the radiation produced by sources **1**, **2** and **3**, and explain your choices. [6]

....

....

....

....

....

(b) How could the conclusions from this experiment be made more reliable? [1]

....

[Total: .... / 7]

1. Label the following diagrams and number them 1–4 in order of size, starting with the smallest. The first one has already been done for you. [4]

....	Comet	....	....
☐	1	☐	☐

2. What is a light year? [1]

....

3. (a) The temperature on the surface of Venus is 467°C and its atmosphere consists of carbon dioxide and nitrogen. Would it be better to send a manned spacecraft or an unmanned probe to explore Venus? Explain your answer. [3]

....

....

....

(b) How would the information we got back from Venus be different from information we were able to obtain from the Moon? Explain your answer. [4]

....

....

....

....

4. The Sun is in the centre of our solar system. The eight planets, including Earth, orbit the Sun. In the space below draw Earth's orbit around the Sun. [1]

Sun

[Total: .... / 13]

Higher Tier

5. What generates the centripetal force that causes the Earth's orbital motion? [2]

6. A light year is equivalent to 10 trillion kilometres. Describe a suitable situation in the study of space where scientists would use kilometres and a situation when they would use light years. [2]

[Total: ____ / 4]

1. Asteroids are large rocks that orbit the Sun.

(a) What is the difference between an asteroid and a comet? [2]

....

....

(b) Scientists believe that asteroids have collided with the Earth in the past. What evidence is there to support this idea? [3]

....

....

....

(c) Describe how a collision between two planets could lead to the formation of a single planet and a moon. [2]

....

....

....

2. Astronomers use telescopes to observe the trajectories of NEOs.

(a) What is an NEO? [1]

....

(b) Suggest why scientists need to pay close attention to NEOs. [1]

....

[Total: .... / 9]

Higher Tier

3. Thousands of asteroids orbit the Sun in a large 'belt'. Why don't asteroids join and form new planets? [2]

4. The table below shows a comet's speed at two different distances from the Sun.

Distance from Sun (million km)	Speed of comet (km/s)
5250	0.879
150	54

Explain the above data. [3]

5. What precautions do astronomers take to reduce the risks from NEOs? [3]

6. What evidence is there that the Earth–Moon system is the result of planetary collision? [3]

[Total: .... / 11]

1. The Big Bang Theory is a theory about how the universe reached its current state.

(a) What are the **two** main things that this theory tells us about the formation of the universe? [2]

....

....

....

(b) Explain how observations of space support this theory. [3]

....

....

....

2. All stars begin life as a huge cloud of gas.

(a) What force causes the gas cloud to collapse and form a star? [1]

....

(b) Which gas do stars initially use? .... [1]

3. This question is about **heavy-weight** stars. Look at the diagram below. Fill in the missing information. [3]

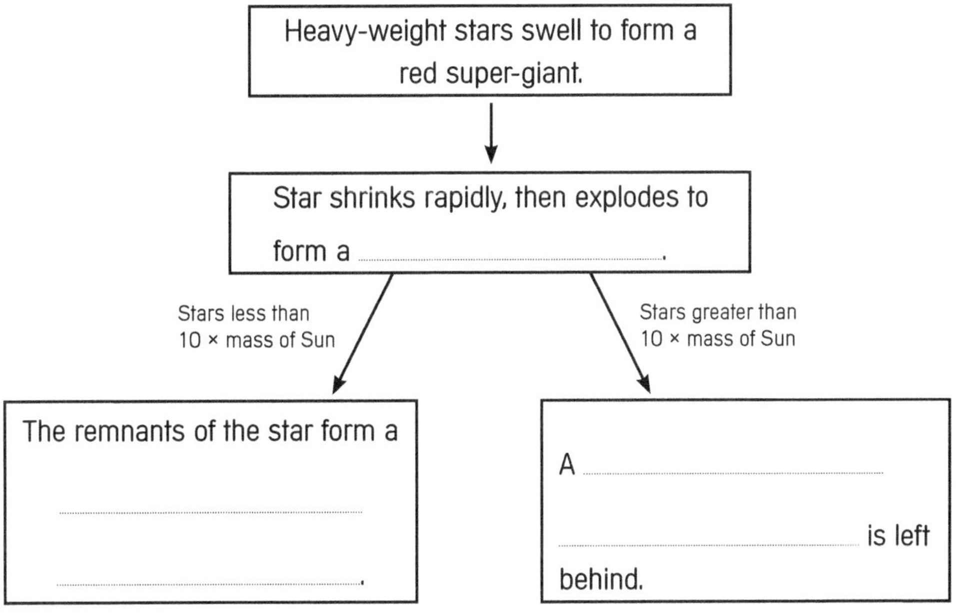

4. This question is about different models of the universe. Look at the following information comparing the Ptolemaic model and the later Copernican model.

Ptolemaic model
• The Earth was the centre of the universe. • The planets and stars were in fixed positions held in place on crystal spheres.

Copernican model
• The Sun was the centre of the universe. • The planets and stars were in fixed positions in the heavens.

(a) Write down **two** reasons why the Copernican model was not accepted for many years. [2]

(b) Galileo supported much of Copernicus' theory, but he believed that the planets were not in fixed positions. What led him to this assumption? [2]

[Total: .... / 14]

Higher Tier

5. 'Red Shift' is evidence that scientists use to support the Big Bang Theory.

(a) What is meant by **red shift**? [2]

(b) How does the light from galaxies close to us differ from light from very distant galaxies? [2]

(c) As well as supporting the Big Bang Theory, what else can scientists use red-shift data to predict? [1]

[Total: .... / 5]

1. What **two** quantities do you need in order to calculate the speed of a moving object? [2]

2. Calculate the average speed of a cyclist who cycles a 3000 m race in $2\frac{1}{2}$ minutes (150 seconds). Use the equations on page 100 to help you. [2]

3. The distance–time graph shows data for Hannah cycling down a road.

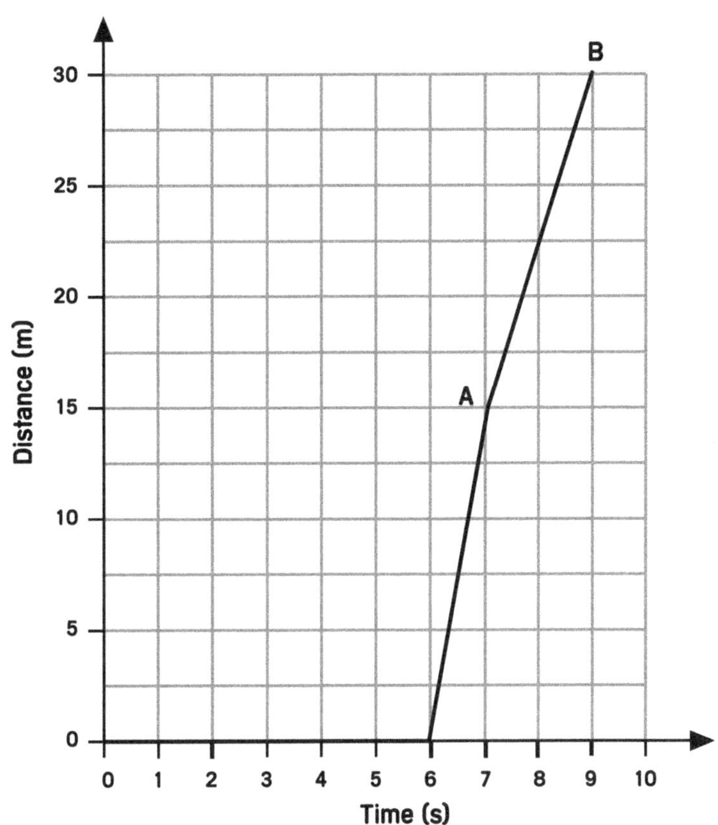

Describe what was happening during [3]

0-6 seconds

6-7 seconds

7-9 seconds

[Total: .... / 7]

Higher Tier

4. The speed of any moving object can be calculated using the equation on page 100. Why is it more appropriate to call the speed calculated the **average speed**? [1]

....

5. **(a)** Calculate the average speed in m/s of a car which travels 6 km in five minutes. [2]

....

....

....

(b) Assuming that the car in part **(a)** continues at the same speed, calculate how long it would take it to travel 50 km. [2]

....

....

6. Study the distance–time graph.

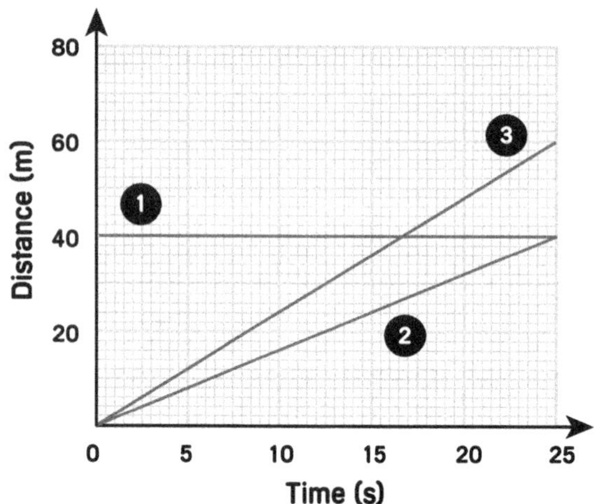

Use the data from the above distance–time graph to calculate the speed of object 2 and object 3. [2]

Object 2 ....

....

Object 3 ....

....

[Total: .... / 7]

1. What **two** quantities do you need to know to calculate the acceleration of an object? [1]

.... **and** ....

2. Two cyclists start from rest. After 5 seconds the speed of the cyclists was measured.

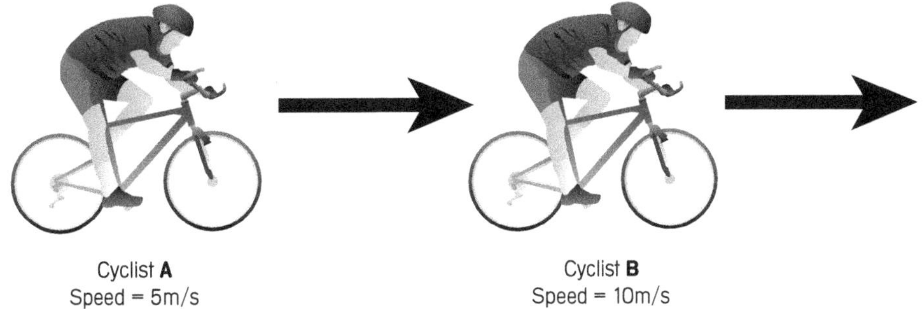

(a) Calculate the acceleration of cyclist **A**. [3]

....

....

....

(b) Without calculation, state which cyclist has a faster acceleration and explain your answer. [1]

....

3. Two cyclists started from the same point. The first cyclist had negative velocity compared to the second one. What does this show? [1]

....

4. Look at the speed–time graph of a car below.

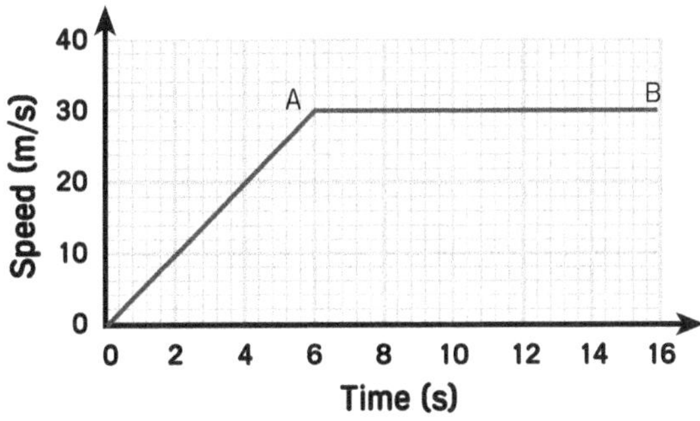

(a) For how long is the car accelerating? [1]

....

(b) After 16 seconds the car decelerated at a uniform rate. How would this appear on the graph? [2]

....

....

5. What does the area under a speed–time graph represent? [1]

....

[Total: .... / 10]

Higher Tier

6. An apple falls from a tree with an acceleration of 10 m/s². It reaches a speed of 4 m/s just before hitting the ground. Calculate how long it took for the apple to fall. [3]

....

....

....

7. Study the speed–time graph.

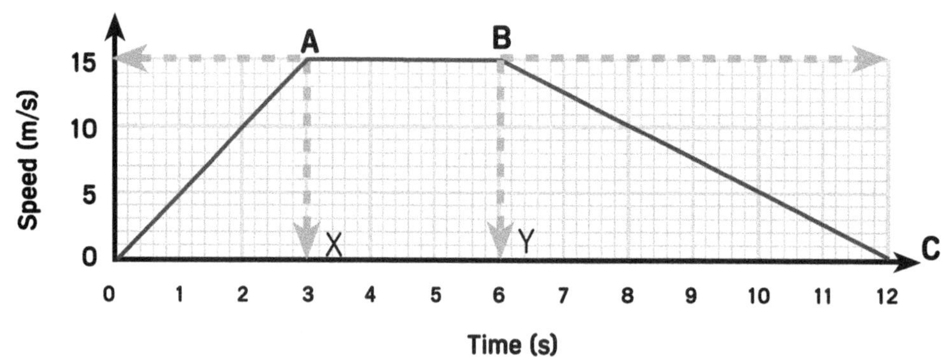

(a) Calculate the rate of acceleration between points **O** and **A**. [3]

....

....

....

(b) Calculate the distance travelled between 0 and 6 seconds. [3]

....

....

....

[Total: .... / 9]

1. A ball of mass 1.5 kg is accelerated at a rate of 2 m/s^2. Calculate the force used to accelerate the ball. [2]

2. Lorry A – Fully loaded   Lorry B – Empty

Two identical lorries accelerate away from traffic lights. They use the same accelerating force from the engine, but lorry **A** contains a full load. Which lorry will accelerate the fastest? Explain your answer. [2]

3. This question is about stopping distances. Look at the picture below of three cars travelling along a road.

A B C

What is represented by [3]

A ....

B ....

C ....

4. A group of scientists were designing an experiment into the effect of different factors on stopping distance. They came up with the following list.

Tiredness

Ice on the roads

Alcohol intake

Wear on brakes

(a) Which of these factors affect **braking** distance and which affect **thinking** distance? [2]

(b) Why is tiredness not a very 'good' variable? [1]

[Total: .... / 10]

Higher Tier

5. A roller coaster car is accelerated from rest to 6 m/s in 2 seconds. The mass of the car and passengers is 900 kg.

(a) Calculate the acceleration of the car. [2]

(b) Calculate the accelerating force. [2]

(c) During testing the roller coaster car was accelerated with the same force. The car had a mass of 600 kg without passengers. Calculate the acceleration of the empty car. [2]

6. Claire was driving her car and her car's temperature gauge showed that it was −9°C outside. She was finding it harder to stop her car quickly. Explain why icy roads increase stopping distance. [3]

7. Describe the differing effects of a doubling of speed on thinking and braking distances. [2]

[Total: .... / 11]

1. Energy is needed to do work. What is the unit of work and energy? .... [1]

2. Jill lifts a box. What **two** factors will affect how much work is done? [1]

.... **and** ....

3. The picture shows a motor lifting a load.

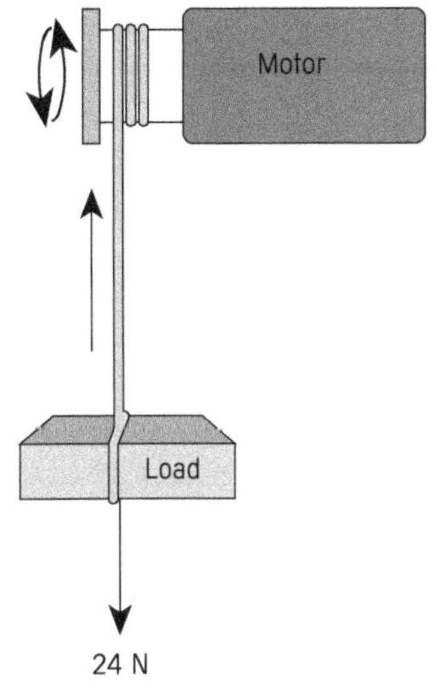

(a) How much work is done when the load is lifted through a height of 0.5 m? [2]

....

....

(b) How much energy is transferred to the load? [1]

....

(c) It takes 3 seconds to lift the load through 0.5 m. Calculate the power of the motor. [2]

....

....

(d) A different load had a mass of 10 kg. Calculate the weight of this load. Assume a gravitational field strength of 10 m/s^2. [1]

....

(e) How high would this load have to be lifted to give the same work done as the previous load? [2]

....

....

4. Nahid was thinking of buying a new car. She made a table of data for fuel consumption and engine size.

Car	Fuel consumption in kilometres per litre	Engine size in cc	CO_2 emissions in grams/kilometre
A	28.0	999	109
B	19.4	1200	142
C	15.2	1449	179

(a) Use the table to explain which car would be the most **environmentally friendly** model. [3]

(b) Assuming all the cars cost the same to buy, is the car you named above the most economical to drive? Explain your answer. [2]

[Total: .... / 15]

Higher Tier

5. A lift carries passengers with a total weight of 3000 N. The lift weighs 5000 N.

(a) What is the total mass of the lift and passengers? (acceleration due to gravity = 10 m/s^2) [2]

(b) Calculate the work done to raise the lift 6 m. [1]

(c) It takes 10 seconds to raise the lift 6 m. Calculate the power required to do this. [1]

[Total: .... / 4]

1. What **two** quantities does the kinetic energy of an object depend on? [1]

.... **and** ....

2. The picture shows two vehicles travelling at 30 metres per second.

5000 kg

1000 kg

Calculate the kinetic energy of the lorry and the car. [3]

....

....

3. Cars and lorries experience drag as they move.

(a) How is the drag experienced by a lorry reduced? [1]

....

(b) A car is travelling with its windows open. How will this affect the fuel consumption of the car? [1]

....

4. 'Electric cars do not cause any pollution.' Explain why this statement is not true. [2]

....

....

5. Why may we have to rely on biofuelled and solar-powered cars in the future? [3]

....

....

....

[Total: .... / 11]

Higher Tier

6. Fuel consumption figures are given as averages by car companies. Explain factors that may cause the actual fuel consumption to differ from these averages. [6]

The quality of your written communication will be assessed in this question.

7. A car moves at a speed of 12 m/s and has a kinetic energy of 72000 J. Calculate its mass. [2]

8. A motorcyclist accelerates on the motorway. He doubles his speed. By how much does his kinetic energy increase? Explain your answer. [2]

[Total: ______ / 10]

1. Modern cars like the one shown below have features that make them safer than older cars during an accident.

Write down **two** features of modern cars which help to **prevent** accidents. [2]

.... **and** ....

2. A moving car has kinetic energy. Describe how safety features such as air bags and crumple zones help during a crash. [3]

....

....

....

3. This question is about momentum. A car in a crash test has a mass of 1000 kg. The dummy driver and passengers have a total mass of 200 kg. The car travels at a velocity of 20 m/s.

(a) Calculate the total mass of the car and its occupants. [1]

....

....

(b) Calculate the total momentum of the car and its occupants. [2]

....

....

.... **kgm/s**

(c) When the car is crashed its momentum becomes zero in 0.5 seconds. Calculate the force generated by the crash. [2]

....

....

[Total: .... / 10]

Crumple Zones

Higher Tier

4. Anti-lock braking systems help cars to stop more safely. Describe how. [3]

....

....

....

5. Modern cars have crumple zones.

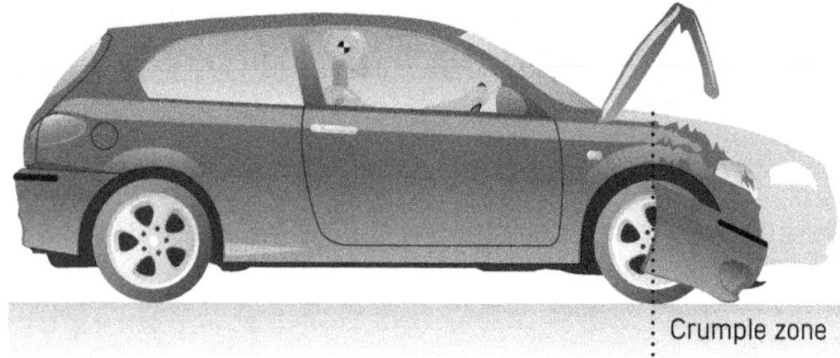

Describe how crumple zones help reduce injury during a collision. [4]

....

....

....

....

[Total: .... / 7]

1. The picture shows a skydiver.

(a) **X** and **Y** are two forces acting on the skydiver as she falls. Name these **two** forces. [2]

X .... **Y** ....

(b) What is the name of the force which causes the skydiver to accelerate? [1]

....

(c) What happens to the size of force **X** as the skydiver accelerates? [1]

....

(d) After a while the skydiver stops accelerating. She continues at a constant speed.

(i) Give the correct name for this 'constant speed'. [1]

....

(ii) The skydiver has not opened her parachute. Explain why her speed is now constant. [2]

....

....

2. The graph shows the speed of a skydiver as he descends through the air.

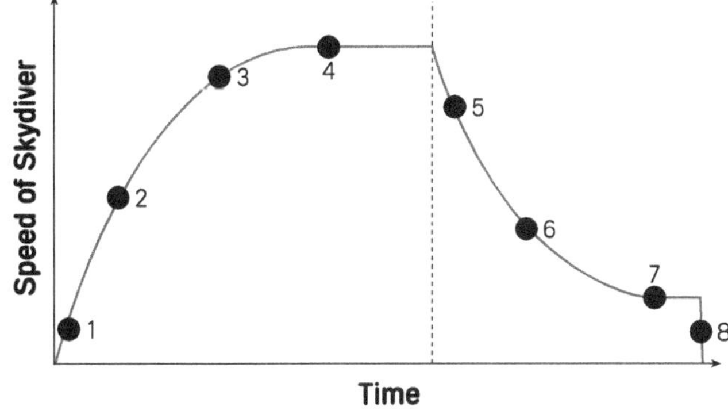

(a) Between which **two** points does the skydiver open his parachute? [1]

Between .... **and** ....

(b) Explain why the skydiver has a lower terminal speed when his parachute is open. [3]

....

....

....

[Total: .... / 11]

Higher Tier

3. A skydiver has a mass of 70 kg. Assume that the acceleration due to gravity is 10 m/s^2.

(a) Calculate the skydiver's weight. [1]

....

....

(b) What will be the value of the air resistance when the skydiver is falling at terminal velocity? Explain your answer. [2]

....

....

(c) Give **two** factors that cause slight variations in acceleration due to gravity. [2]

....

....

[Total: .... / 5]

1. The diagram shows a skydiver.

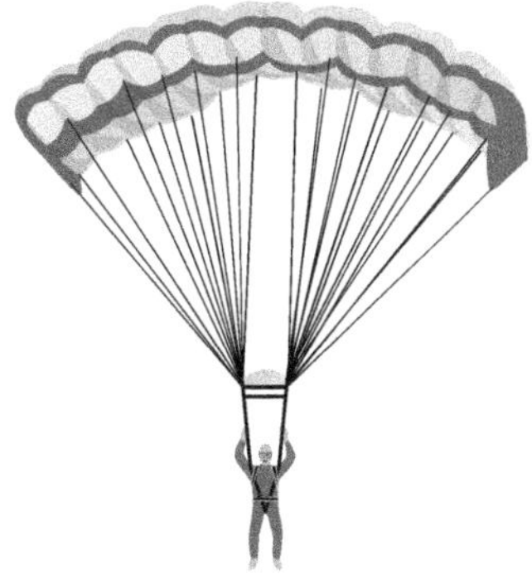

(a) What kind of potential energy does the skydiver have? [1]

(b) What **two** factors does the amount of GPE depend on? [2]

(c) What will happen to the amount of GPE the skydiver has as he falls? [1]

2. (a) Astronauts landed on the moon. They dropped a feather and a hammer from the same height. Which object had the greatest gravitational potential energy? Explain your answer. [2]

(b) The hammer has a mass of 2 kg and was dropped from a height of 1.2 metres. The acceleration due to gravity on the moon is 1.6 m/s^2. Calculate the hammer's gravitational potential energy just before it is dropped. [2]

3. The picture shows a roller coaster.

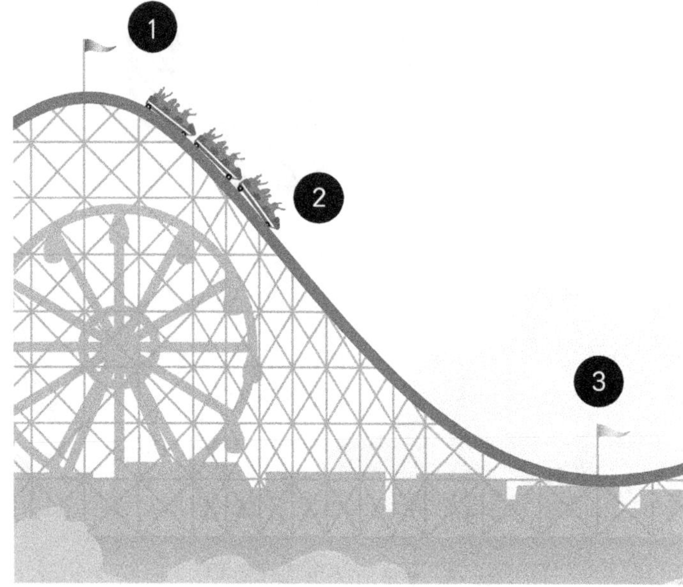

(a) What type of energy does the roller coaster have when it is at point **1**? [1]

....

(b) What does this energy transform into as the roller coaster travels down the slope? [1]

....

(c) At which point does the roller coaster have the least gravitational potential energy? [1]

....

(d) What would be the effect on kinetic energy of doubling the mass of the roller coaster? [1]

....

[Total: .... / 12]

Higher Tier

4. A box sits on a shelf 3.5 m above the ground. It has a gravitational potential energy of 490 J. Use $g = 10\ m/s^2$ to calculate its mass. [2]

5. Richard jumps out of a plane. His speed increases and then he reaches a terminal velocity.

Describe the energy transfers occurring as Richard jumps from the plane and then reaches terminal velocity. [4]

6. John and Sushant are designing a roller coaster. They want to increase the maximum kinetic energy of the cars. Should they double the mass of the cars or the speed of the cars? Explain your answer. [2]

7. A cyclist climbs from a height of 100 m to a height of 300 m. The cyclist has a mass of 70 kg. By how much does the cyclist's gravitational potential energy increase? [2]

[Total: .... / 10]

P4 Sparks

1. This question is about static electricity. Complete the following sentences. Choose from the words below. [1]

electrons **attract** **conducting** **repel** **insulating** **protons**

insulating materials can become charged when rubbed with a duster.

electrons are transferred from one material to the other.

Oppositely charged materials will .... one another.

2. Joe is wearing rubber-soled shoes. He walks across some carpet. When he touches a metal tap he feels a shock. Explain why. [4]

3. Static electricity can be a nuisance. It can also be dangerous. Give **two** examples of static electricity being a nuisance and **two** examples of it being dangerous. [4]

Nuisance ....

Dangerous ....

4. Tom is writing about electrostatic charge. Look at his notes. Put a tick (✓) next to each of the statements that are correct and a cross (✗) next to each of the statements that are wrong. One has been done for you. [4]

All materials become charged when rubbed with a cloth. ☐

Friction causes electrons to move from one material to another. ☐

A positively charged Perspex rod will repel a negatively charged piece of paper. ☐

Two oppositely charged objects will attract one another. ✓

When hair is combed the hair may stick up because each hair has the same charge. ☐

[Total: .... / 13]

Higher Tier

5. When a Perspex rod is rubbed with a cloth it becomes positively charged.

   (a) Why does the Perspex rod become positively charged? [1]

   ....

   (b) What charge will the cloth have after it is used to rub the rod and why? [2]

   ....

6. Describe what is meant by a **negative ion**. [2]

   ....

   ....

7. The picture shows fuel being pumped into an aircraft. When fuel is being pumped into an aircraft there is a risk of explosion.

   (a) Explain why. [4]

   ....

   ....

   ....

   ....

   (b) How is the risk of explosion reduced? [1]

   ....

[Total: .... / 10]

P4 Uses of Electrostatics

1. State **two** uses of electrostatics. [2]

2. The picture shows a car being spray painted.

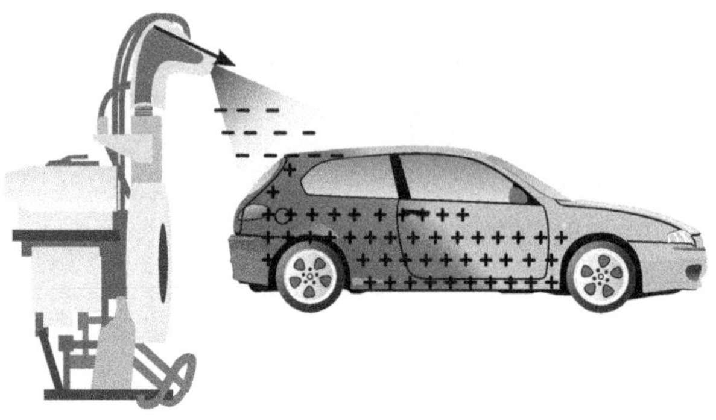

Explain how electrostatics enables a car to be painted without any drips. [4]

3. Mary uses a defibrillator to restart a patient's heart.

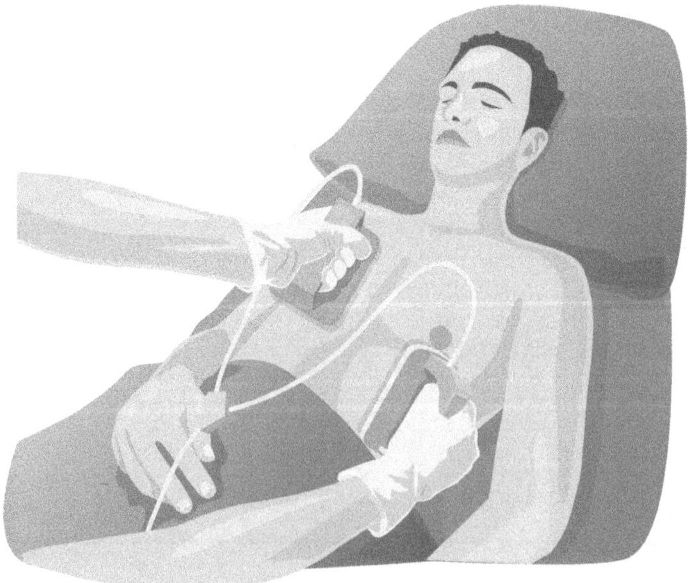

(a) What happens to the heart when a charge is passed through it? [1]

(b) What precautions must Mary take when using a defibrillator? Explain your answer. [3]

[Total: .... / 10]

Higher Tier

4. Look at this picture of a smoke precipitator.

Smoke Precipitator

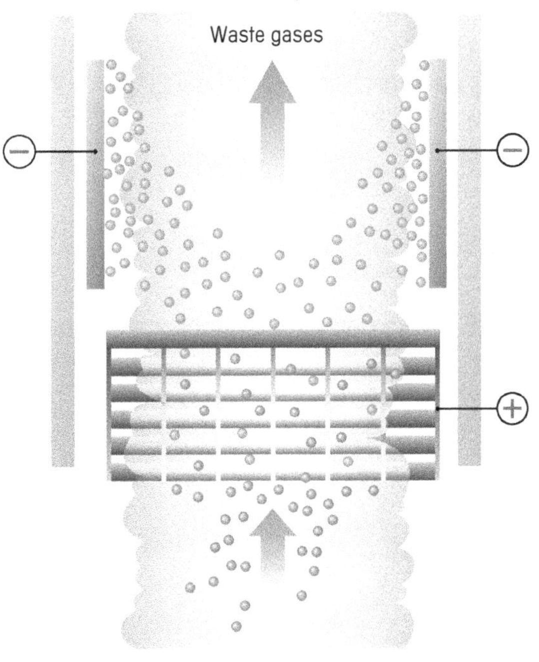

Smoke and waste gases

Explain, using the concept of gaining and losing electrons, how the smoke precipitator carries out its function. [3]

[Total: .... / 3]

P4 Safe Electricals

1. This question is about resistance in a circuit.

   (a) What is the effect on current of increasing the resistance of a circuit? [1]

   ....

   (b) The unit of voltage is the volt. What is the unit used to measure resistance? [1]

   ....

   (c) What would be the effect of increasing the voltage through a resistor? [1]

   ....

   (d) If the voltage remained high but the resistance was increased, what would be the effect on the current? [1]

   ....

2. A simple circuit allows 0.5 A to flow through it when a voltage of 200 V is applied across it. Calculate the resistance of the circuit. [2]

   ....

   ....

3. Look at the diagram of a plug. Give the names and functions of the two wires shown. [4]

   A
   B
   5A

   A ....

   ....

   B ....

   ....

4. Some appliances are **double insulated**.

   How are double insulated appliances different from other appliances? [2]

   ....

   ....

5. A fuse is a safety device found in a plug. Describe how a fuse protects the user. [4]

6. This question is about electrical power.

A vacuum cleaner is plugged into the mains (230 V). A current of 11.3 A flows. Calculate the power rating of the vacuum cleaner. [2]

[Total: .... / 18]

Higher Tier

7. A student was investigating the resistance of wires of different materials. His results are shown in the table below.

Material	**Resistance/ohms**
Copper	0.02
Steel	0.1

(a) When using the copper wire, the student measured a constant voltage of 0.2 V in the experiment. Calculate the current used in the experiment. [2]

(b) Many appliances have a metal casing. Describe how the earth wire and fuse work together to protect the user from shock if a fault occurs. [4]

[Total: .... / 6]

P4 Ultrasound

1. What type of wave is **ultrasound**? [1]

....

2. **(a)** What is meant by a longitudinal wave's [2]

   Frequency ....

   Wavelength ....

   (b) Distinguish between compression and rarefaction in longitudinal waves. [2]

   ....

   ....

3. Why can't humans hear ultrasound? [2]

....

....

4. An ultrasound pulse from a boat is directed at the sea floor. Ultrasound travels at 1400m/s. The pulse takes 0.5 seconds to return to the boat. What is the depth of the sea in this area? [3]

....

....

....

5. The speed of flow of which bodily fluid can be measured using ultrasound? [1]

....

6. Apart from scanning, give **one other** use of ultrasound in medicine. [1]

....

[Total: .... / 12]

Higher Tier

7. The picture shows an ultrasound scan.

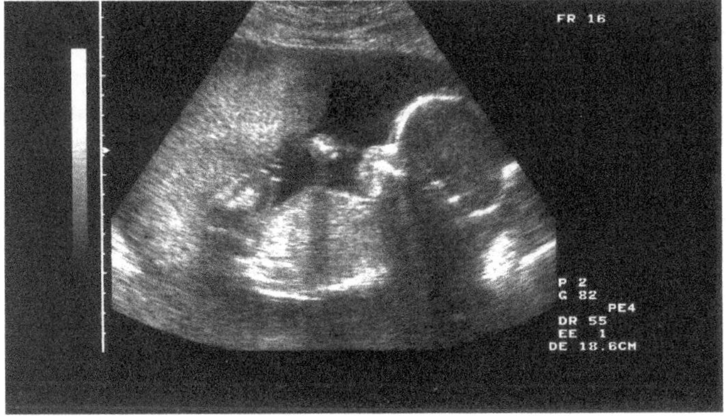

Explain how ultrasound is used to observe an unborn foetus. [4]

....

....

....

....

8. What are the **advantages** of using ultrasound instead of X-rays to view an unborn foetus? [2]

....

....

[Total: .... / 6]

P4 What is Radioactivity?

1. How is the radioactivity of a substance measured? [2]

2. (a) What **three** kinds of radioactivity are produced by the natural decay of a radioactive substance? [3]

   (b) In an experiment into radioactive decay, two types of particle were emitted from a radioactive source. Two types of radiation involve a particle being emitted; what is the difference between these two particles? [2]

3. What is meant by the half-life of a substance? [1]

[Total: .... / 8]

Higher Tier

4. Use the periodic table on page 99 to answer the following questions.

   (a) $^{232}_{90}Th$ decays by alpha emission. Write a **balanced** equation to show this decay. [3]

   (b) An isotope of actinium (Ac) has a mass number of 228. Actinium-228 decays by beta emission. Determine the new element produced when actinium decays, and write a balanced equation to show this decay. [4]

5. The activity of a sample of radioactive material was recorded over an eight-hour period.

Time/hours	0	1	2	3	4	5	6	7	8
Activity/Bq	400	290	205	140	95	70	50	35	20

(a) Plot the data on the axes. [1]

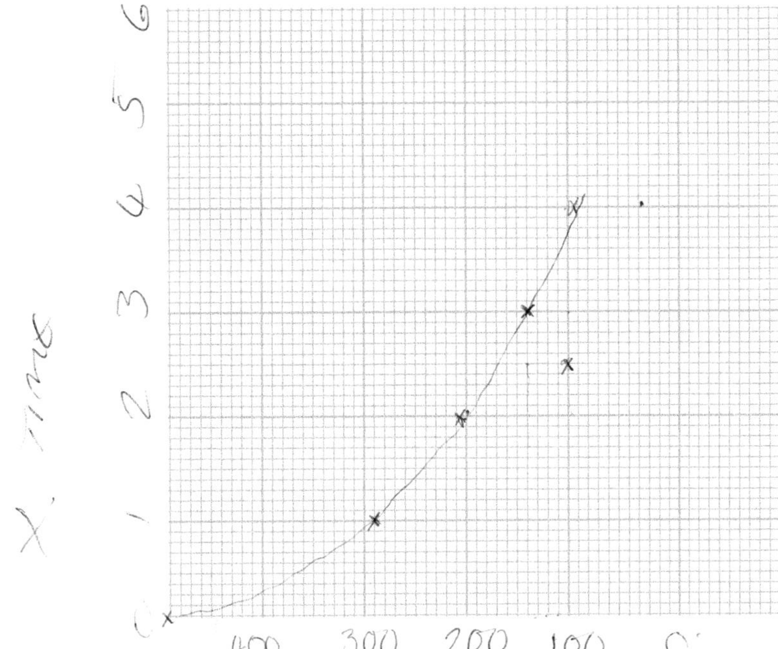

(b) Draw a curve of best fit. [1]

(c) Use the graph to determine the half-life of the substance. [2]

....

....

[Total: .... / 11]

P4 Uses of Radioisotopes

1. What are the **two** main natural contributors to background radiation? [2]

.... **and** ....

2. Explain how a radioactive tracer could be used to check for a blockage in a pipe. [3]

....

....

....

3. This is a diagram of a smoke detector.

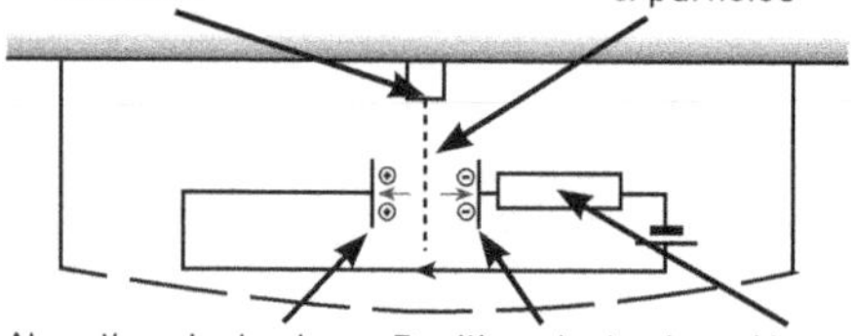

Explain the purpose of the alpha emitter in the smoke detector. [4]

....

....

....

....

[Total: .... / 9]

Higher Tier

4. Explain how carbon dating is used to age ancient artefacts. [4]

....

....

....

....

[Total: .... / 4]

1. Describe **two similarities** and **one difference** between X-rays and gamma rays. [3]

Similarities ....

....

Difference ....

2. Technicium-99 has a half-life of six hours. Explain why Technicium-99 can be used as a medical tracer but Uranium-235 cannot. [3]

....

....

....

....

[Total: .... / 6]

Higher Tier

3. Gamma rays can be used to treat tumours. Explain why gamma rays can be used to treat cancer and describe some of the potential risks and benefits of this treatment. [6]

The quality of your written communication will be assessed in this question.

....

....

....

....

....

....

....

....

....

....

[Total: .... / 6]

Fission and Fusion

1. Domestic electricity can be produced in a nuclear power station. Describe the main stages of electricity production. [4]

2. Nuclear fission and nuclear fusion release energy.

   (a) What happens during nuclear fission? [1]

   (b) Why can't nuclear fusion be used to produce power efficiently at the moment? [2]

3. One group of scientists claims to have successfully carried out a cold fusion experiment. Explain why it is important that the experimental method and any data for this experiment is shared amongst the scientific community. [3]

[Total: .... / 10]

Higher Tier

4. A neutron is fired into a uranium nucleus to start a chain reaction inside the core of a nuclear power station.

   (a) Explain how a chain reaction continues once started. [3]

   (b) Control rods are used in a nuclear core to prevent the chain reaction getting out of control. How do control rods stop the chain reaction getting out of control? [2]

[Total: .... / 5]

Satellites, Gravity and Circular Motion

1. (a) What is a satellite? [1]

(b) Give the name of the force that keeps a satellite in orbit. [1]

(c) Describe the difference between an **artificial** and a **natural** satellite. [2]

2. Describe the orbit of a geostationary satellite. [2]

3. Satellites in low polar orbit can be used by the military for spying. Geostationary satellites cannot. Explain why. [3]

4. What force is required to give the circular motion of a spy satellite? What provides this force? [2]

[Total: .... / 11]

Higher Tier

5. The diagram shows a comet.

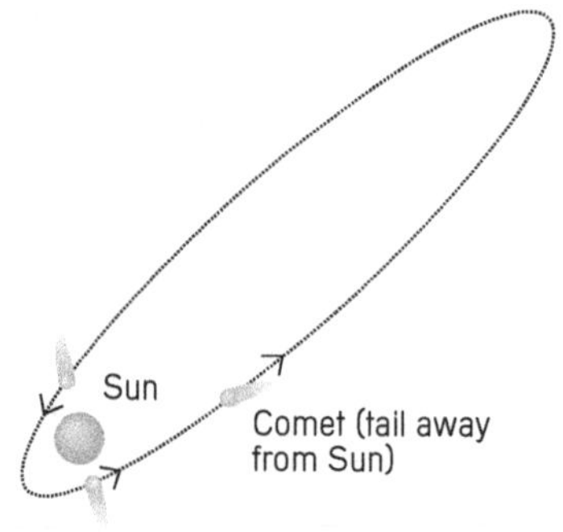

Sun

Comet (tail away from Sun)

Describe and explain the motion of the comet. [6]

✏ *The quality of your written communication will be assessed in this question.*

....

....

....

....

....

....

....

....

....

....

[Total: .... / 6]

1. This question is about scalars and vectors. What is the difference between vector and scalar quantities? Give an example of each. [4]

2. Study the picture of two cars.

Car A

Car B

Car **A** moves to the right at a speed of 6 metres per second. Car **B** moves to the left at a speed of 8 metres per second.

(a) Calculate the relative speed at which the two cars approach one another. [2]

14 m/s

(b) Car **B** then turns around and travels at 8 metres per second in the same direction as car **A** at 6 m/s. What would the relative speed be now? [2]

3. **(a)** A ball falls off the edge of a cliff (u = 0 m/s). It accelerates at a rate of 10 m/s^2 for 3 seconds before hitting the ground. Calculate the final velocity of the ball. [3]

30 m/s.

(b) Calculate the height of the cliff. [2]

(c) A ball is thrown off a different cliff of height 60 m. Its initial downward velocity is 5 m/s and its final velocity is 35 m/s. How long does it take for the ball to hit the ground? [2]

[Total: / 15]

Higher Tier

4. Sian is driving a car. The car accelerates from rest at a rate of 4 m/s^2 for 50 metres. Calculate the car's final velocity. [3]

4

5. A ball is fired vertically upwards with an initial velocity of 20 m/s. The acceleration due to gravity is 10 m/s^2. Remember that final velocity at maximum height = 0. Calculate the maximum height the ball reaches. [4]

[Total: / 7]

1. What is the path of a projectile called? [1]

2. Sarah is trying to kick a football as far as possible. She predicts that if she keeps increasing the angle above the ground that the range of the ball will keep increasing.

The table shows the data she collected.

Angle in degrees	Maximum height in m	Range in m
10	4	25
20	18	54
30	35	69
40	50	75
50	73	67
60	87	53

Use the data and your scientific knowledge to explain whether Sarah's prediction was correct. [3]

[Total: / 4]

Higher Tier

3. Louise kicks a stone off the edge of a cliff. Its initial horizontal velocity is 7 metres per second. Its initial vertical velocity is zero.

(a) The stone takes 4 seconds to hit the beach below. Calculate the height of the cliff. [3]

(b) What is the horizontal range of this stone? [3]

4. Adam is rowing a small boat across a river. He is trying to row from point **A** to point **B**, but the force from the water is causing him to head downstream.

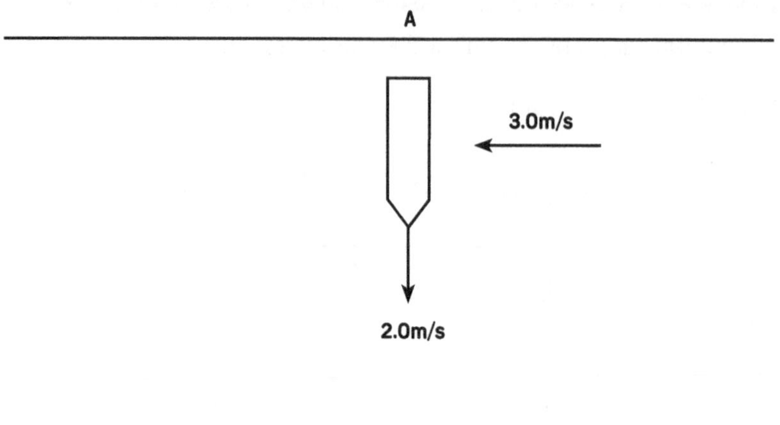

Calculate the size of the resultant velocity of the boat. [3]

5. A tennis player hits a ball horizontally. Describe the effect of gravity on its horizontal and vertical velocity. [2]

[Total: .... / 11]

1. Fill in the blanks below.

(a) Every action has an .... and .... reaction. [2]

(b) If aerosol cans are heated they explode. Use the concept of particles to explain why this occurs. [3]

(c) How does burning rocket fuel cause a rocket to lift off? [3]

[Total: .... / 8]

Higher Tier

2. **(a)** A soldier fires a gun in training and his arm is bruised by the recoil of the gun. Explain what is meant by recoil and why it occurs in terms of momentum. [4]

(b) Robin is an Arctic scuba diver. He notices that the pressure inside his air cylinder decreases when he takes it from the heated boat cabin out into the Arctic air. Explain this effect. [3]

3. Two cars are involved in a head-on collision. Car **A** has a mass of 1100 kg, and car **B** has a mass of 800 kg. Just before the collision they each have a velocity of 10 m/s. During the collision they become joined together.

(a) Calculate the final momentum of the joined cars. [3]

....

....

....

....

(b) Calculate the final velocity of the joined cars. [3]

....

....

....

....

[Total: .... / 13]

1. Franca has a phone with a direct link to a satellite (a sat phone). It can be used in areas where a normal mobile phone would not have a signal.

Explain why a sat phone can be used where a mobile phone cannot. [4]

....

....

....

....

2. **(a)** Describe the behaviour of the following electromagnetic waves in the atmosphere. [3]

Below 30 MHz ....

Above 30 GHz ....

Between 30 MHz and 30 GHz ....

(b) Explain which of the above frequencies would be used by a satellite phone. [2]

....

....

[Total: .... / 9]

Higher Tier

3. Satellites are used for military communication. Explain why satellite transmitting and receiving dishes need very careful alignment. [2]

....

....

4. When a sound wave passes through a doorway it undergoes diffraction. Light undergoes very little diffraction as it passes through the same doorway, and continues in a straight line. Explain why. [4]

....

....

....

....

[Total: .... / 6]

1. This question is about light. Complete the following sentences. Choose from the words below. [1]

vacuum **reflect** **interfere** **straight** **longitudinal**
diffracted **electromagnetic**

Light, microwaves and infrared are all .... waves. When light passes through a tiny gap it can be .... This makes it look as though light bends. When this happens the waves can overlap and ....

2. When waves overlap they can cause interference.

(a) What type of interference is observed when two peaks overlap? [2]

....

....

(b) What would be observed if two peaks of light waves overlapped? [1]

....

3. The experiment shown below can be used to determine the wavelength of light.

(a) What conditions are needed for a stable interference pattern to be produced? [2]

....

....

(b) What would be observed if the double slit was replaced with a single slit? [2]

....

....

4. A wave of light is incident on a horizontally polarising film. Explain how the polarising film will affect the light. [3]

....

....

....

[Total: .... / 11]

Higher Tier

5. For a stable interference pattern the light must be from coherent sources. Explain what this means. [3]

....

....

....

6. Two light waves are emitted from a lighthouse. The two rays interfere after one ray has been reflected off a nearby object. The second ray has effectively travelled 6000 half wavelengths more than the first ray before they interfere.

(a) Is the interference **constructive** or **destructive** at this point? [1]

....

(b) What would be observed at this point? [1]

....

....

7. Light is incident on a horizontal polariser. A second polariser is placed behind the first and is turned through 90°. What will be observed? Why? [4]

....

....

....

....

[Total: .... / 9]

1. This question is about refraction. Complete the following sentences. Choose from the words below. [2]

radiation **towards** **more** **medium** **speed** **less**

A .... is a substance that light can travel through.

When light travels from one substance into another its .... changes.

As light travels into a .... dense substance it refracts ....

the normal.

2. What does the refractive index of a material tell us about light entering that material? [1]

....

3. Light travels at 3.0×10^8 m/s in a vacuum. When light enters the lens of the eye it slows down to 2.2×10^8 m/s. Calculate the refractive index of the lens. [4]

....

....

....

....

4. The diagrams show light hitting a glass-air boundary.

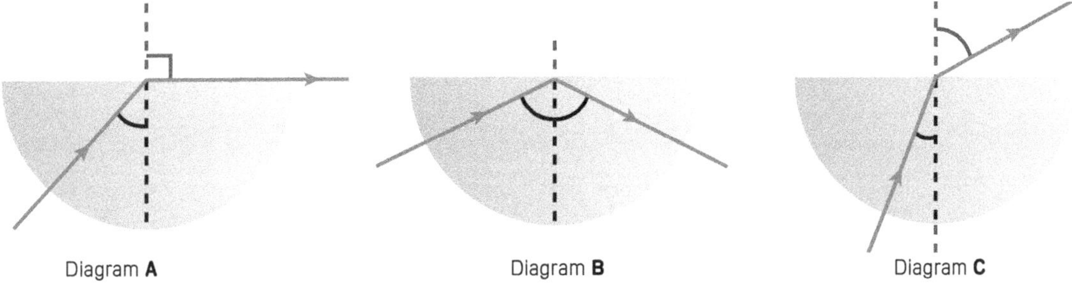

Which diagram shows what happens when light hits a glass-air boundary **at** the critical angle? [1]

....

5. When light travels through a prism it is dispersed, as shown in the diagram below.

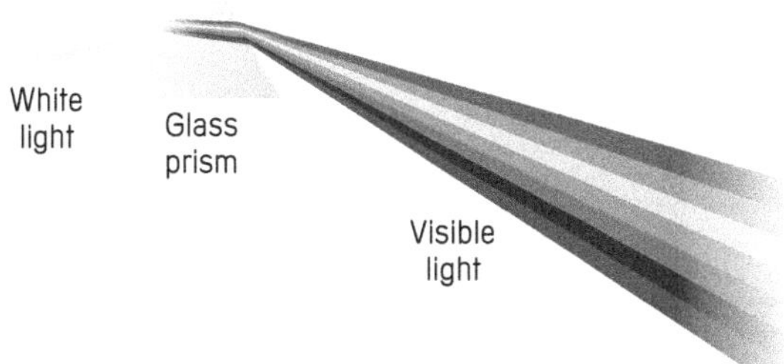

P5 Refraction of Waves

(a) Look at the table about light.

Colour of light	Wavelength in m	Refractive index
Blue	4.30×10^{-7}	1.53
Red	7.04×10^{-7}	1.51

Use the information in the table to explain why the light disperses as it passes through the prism. [2]

....

....

....

(b) Which colour, red or blue, is slowed down the most as it enters the prism? [1]

....

[Total: .... / 11]

Higher Tier

6. The refractive index of diamond is 2.4. Calculate the speed of light in diamond. [3]

(Speed of light in a vacuum = 3.0×10^8 m/s).

....

....

....

[Total: .... / 3]

1. The diagram shows light passing through a lens.

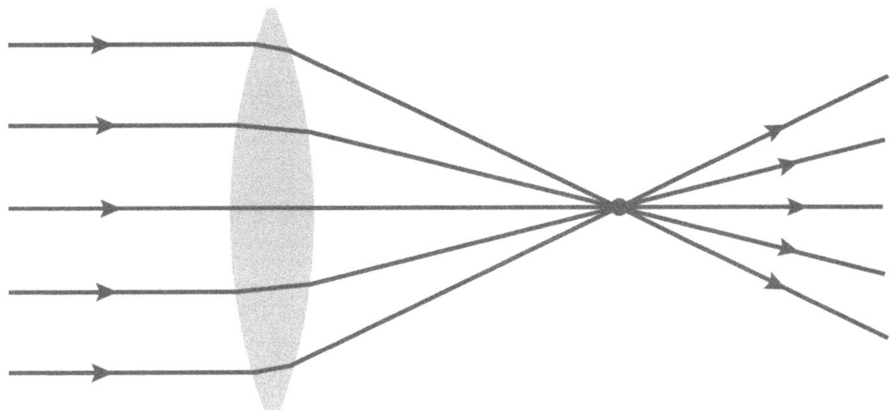

(a) What type of lens is shown in the diagram? [1]

....

(b) Label the following on the diagram: [3]

Focal point

Focal length

Principal axis

2. Give **four** uses of convex lenses. [4]

....

....

....

....

3. Arun used a magnifying glass to look at a centipede which is 1.2 cm long. The image produced was 3 cm long. Calculate the magnification of the lens. [2]

....

....

....

[Total: .... / 10]

Higher Tier

4. A 10m-high image is produced using a magnification of 60. Calculate the height of the original object. [2]

....

....

5. Describe what is meant by **real** and **virtual** images. [4]

Real ....

....

Virtual ....

....

[Total: .... / 6]

1. Jamie sets up the following circuit.

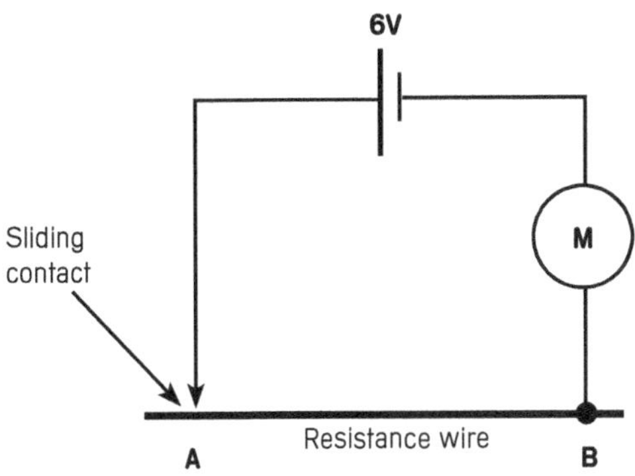

Jamie slides the contact along the resistance wire from **A** to **B**. The speed of the motor changes. How does the speed of the motor change? Explain your answer. [3]

....

....

....

2. A piece of wire is an **ohmic** conductor.

(a) If the voltage across an ohmic conductor is doubled, what will happen to the size of the current? [1]

....

(b) An ohmic conductor has a resistance of 48 ohms. When a current of 0.5 A passes through it, what is the voltage? [2]

....

....

3. Study the graph below.

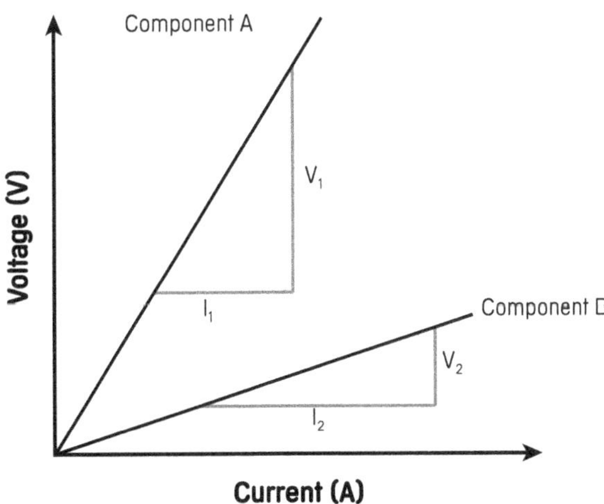

Which component, **A** or **B**, has the greatest resistance? Explain your answer. [2]

....

....

[Total: .... / 8]

Higher Tier

4. William sets up a circuit to observe how the current through a bulb changes as the voltage across the bulb is increased. The graph shows his results.

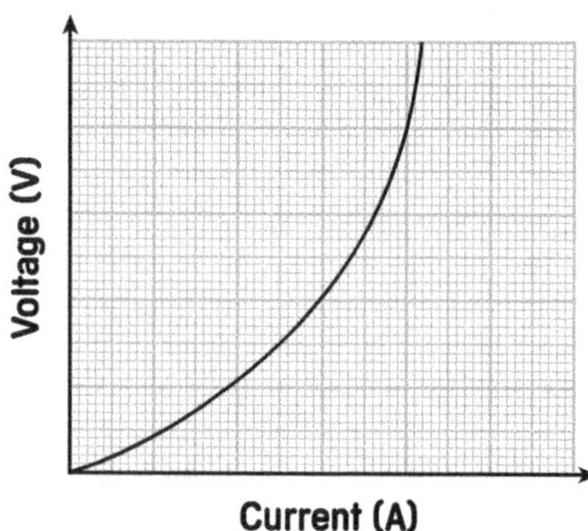

Use kinetic theory to explain why the graph is curved. [6]

The quality of your written communication will be assessed in this question.

....

....

....

....

....

....

....

....

....

....

[Total: .... / 6]

1. This question is about resistors in series and parallel.

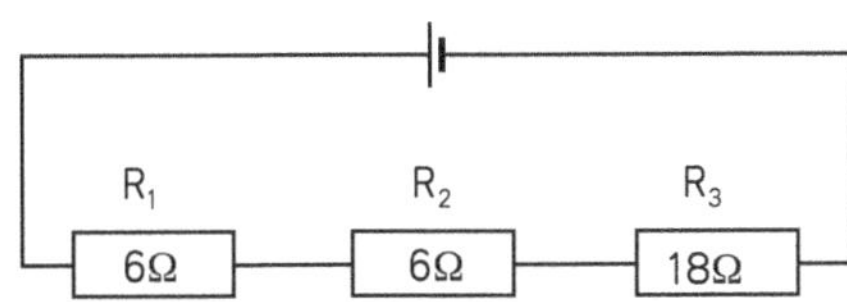

(a) Look at the series circuit in the diagram. Calculate the total resistance of the circuit. [2]

....

....

(b) The circuit is rearranged so that the resistors are now in parallel. Will the total resistance of the circuit **increase**, **decrease** or **remain the same**? [1]

....

2. An LDR was part of the circuit of an outdoor light. How would the resistance of the LDR differ between day and night? [2]

....

....

[Total: .... / 5]

Higher Tier

3. Look at the potential divider circuit in the diagram.

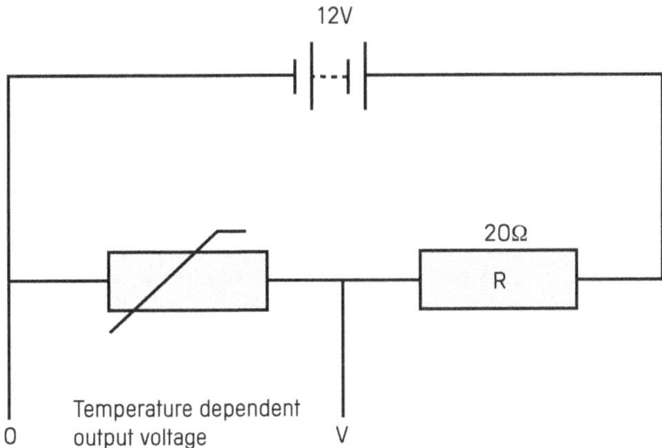

(a) (i) At a particular temperature the resistance of the thermistor is 20 Ω. What will the output voltage across the thermistor be? [1]

....

(ii) The temperature of the thermistor is increased. What happens to the resistance of the thermistor? [1]

....

(iii) What will happen to the output voltage across the thermistor as the temperature increases? [1]

....

(b) At a particular temperature the resistance of the thermistor is 10 Ω. Calculate the output voltage across the thermistor. [3]

....

....

....

(c) Give **one** use for a potential divider circuit which contains an LDR. [1]

....

4. Look at the potential divider circuit.

R
Light
Light dependent output voltage
0
V

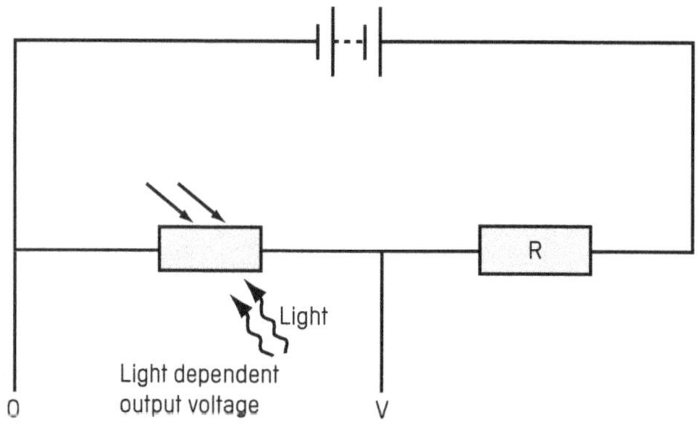

Jenny uses the output voltage to control a buzzer. In what conditions will the buzzer sound? Explain your answer. [3]

....

....

....

....

[Total: .... / 10]

1. **(a)** What is a transistor? [1]

(b) Transistors can be linked to form logic gates. What are the **two** input signals for a logic gate? [2]

2. Complete the truth table for an OR gate. [4]

A	B	Q
0	0	
0	1	
1	0	
1	1	

3. Describe **two benefits** and **two drawbacks** of the miniaturisation of electronic components. [4]

4. In a transistor calculate the current through the emitter if the base current is 0.1 A and the current through the collector is 1.3 A. [2]

[Total: .... / 13]

Higher Tier

5. Complete the following diagram to show the arrangement of transistors in an AND gate. [2]

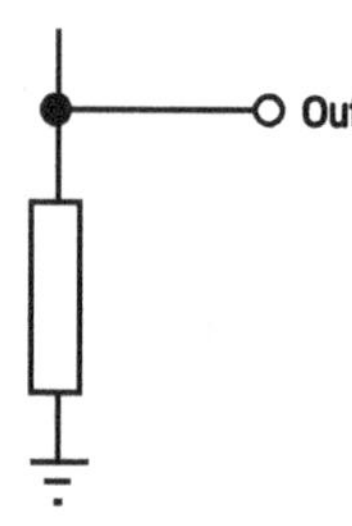

6. Complete the following truth table for a NOR gate. [2]

A	B	Q
0	0	
0	1	
1	0	
1	1	

[Total: .... / 4]

1. This circuit diagram shows an AND gate followed by an OR gate.

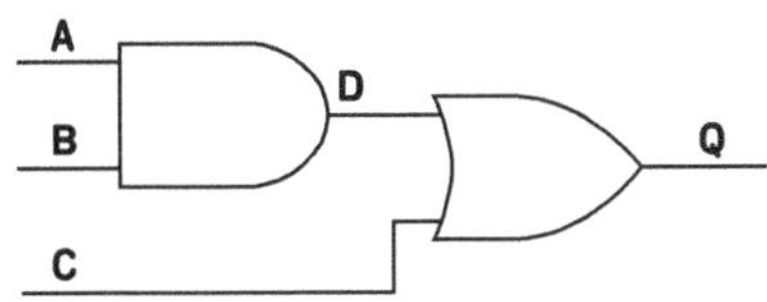

(a) Redraw this circuit, but replace the OR gate with a second AND gate. [1]

(b) Complete the truth table for the circuit containing **two** AND gates. [4]

A	B	C	D	Q
0	0	0	0	
0	0	1	0	
0	1	0	0	
0	1	1	0	
1	0	0	0	
1	0	1	0	
1	1	0	1	
1	1	1	1	

2. Draw the symbol for an LED. [1]

3. The diagram shows a relay circuit. Describe how a relay circuit works. [2]

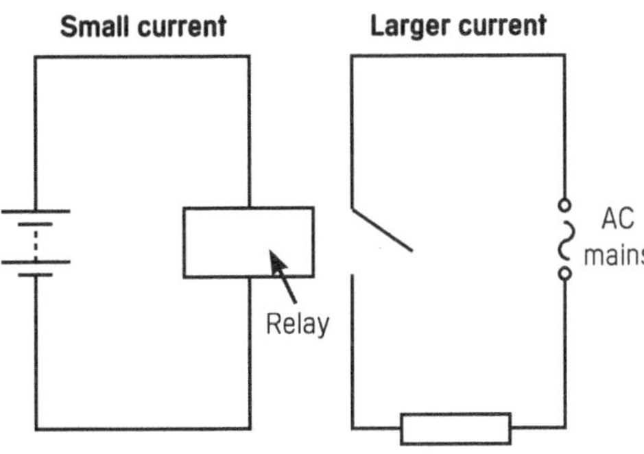

....

....

[Total: .... / 8]

Higher Tier

4. A relay is needed for a logic gate to switch on a current in a mains circuit. Explain why. [3]

5. Study the circuit diagram.

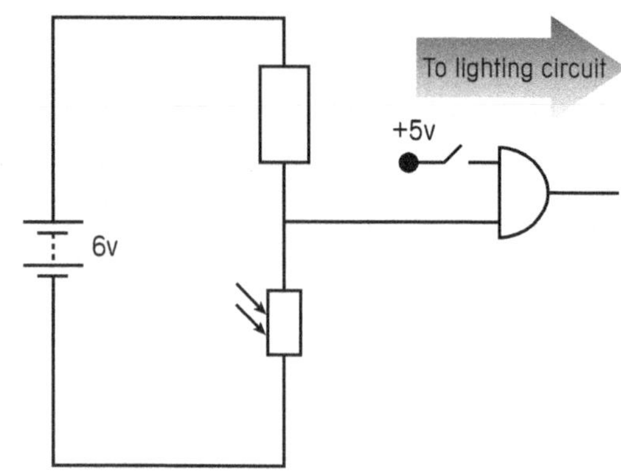

(a) Describe how the circuit in the diagram can be used as a control circuit to switch on the logic gate. [4]

(b) How can the circuit be changed to ensure that the logic gate receives a high input at a particular light intensity? [1]

[Total: .... / 8]

1. When a current is passed through a wire a magnetic field is formed around the wire.

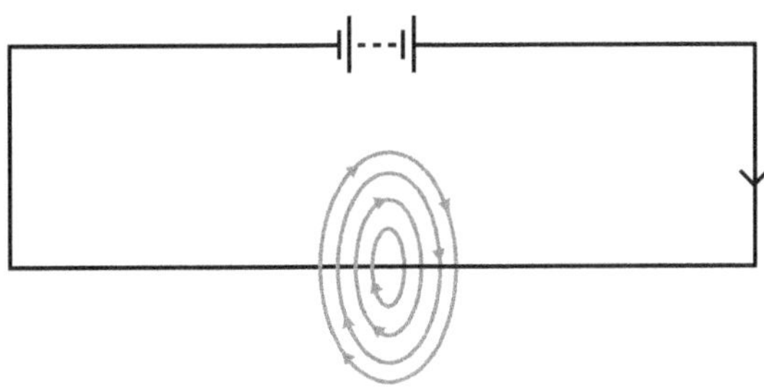

(a) What is the shape of the magnetic field around a coil of wire? [1]

....

(b) Write down the name given to a coil of wire carrying an electric current. [1]

....

(c) The strength of the magnetic field can be increased by adding an iron core inside the coils. Write down **two other** ways to increase the strength of the magnetic field. [2]

....

....

(d) The coil of wire is found to have a north and south pole like a bar magnet. What happens to these poles if the direction of the current is reversed? [1]

....

2. Katie places a wire in a magnetic field. When a current is passed through the wire it moves.

(a) Explain why the wire moves when the current passes through it. [3]

....

....

....

(b) What **two** things could Katie do to make the wire move in the opposite direction? [2]

....

....

(c) How does Katie ensure that the wire experiences the maximum force in the magnetic field? [1]

....

[Total: .... / 11]

Higher Tier

3. Use Fleming's Left-hand Rule to determine the direction of the movement of the wire. [1]

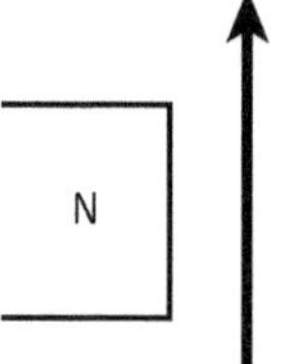

....

....

4. The diagram shows an electric motor.

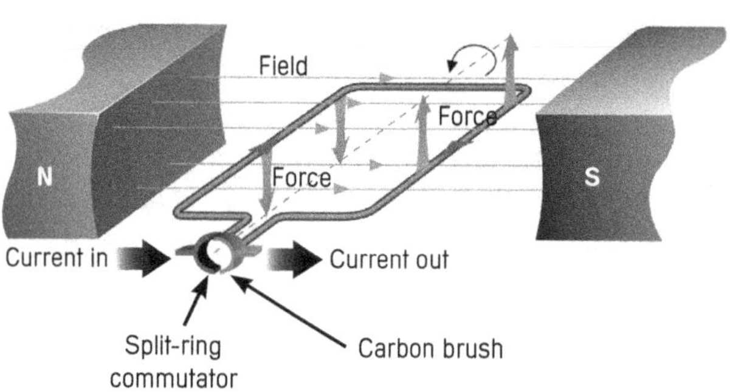

Describe the purpose of the split-ring commutator. [3]

....

....

....

5.

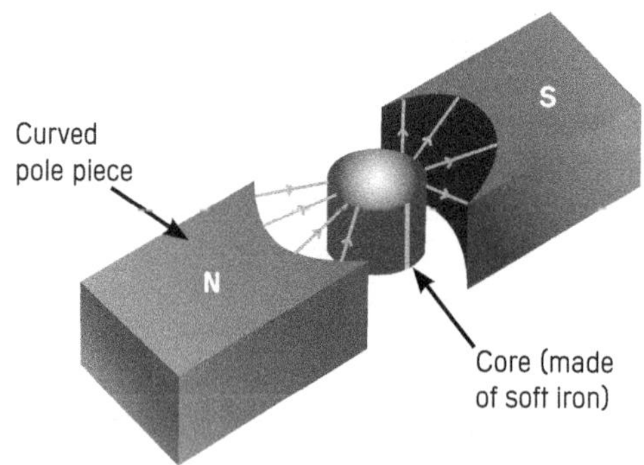

Explain why the magnets in a motor may be curved, as in the diagram above. [3]

....

....

....

[Total: .... / 7]

1. Electricity is generated by moving a magnet near a wire. What is the name of this effect? [1]

....

2. What is the frequency of mains electricity in the UK? [1]

....

3. Using the diagram, explain the effect on the bulb of...

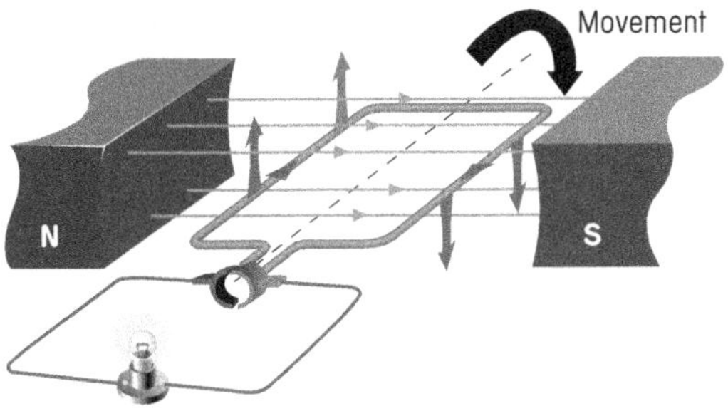

(a) Increasing the speed of rotation. [1]

....

(b) Decreasing the strength of the magnets. [1]

....

(c) Replacing the single coil with ten identical coils. [1]

....

[Total: .... / 5]

Higher Tier

4. In what way does the structure of an AC generator differ from a DC generator? [2]

....

....

5. Explain how and why changing the speed of rotation of the coil within the magnetic field affects the induced voltage produced by an AC generator. [2]

....

....

....

[Total: .... / 4]

P6 Transforming

1. This question is about heat in the National Grid. Complete the following sentences. Choose from the words below. [2]

   step-up **voltage** **heat** **current** **sound** **large** **small** **step-down**

   Overhead power cables lose energy as .... To reduce the amount of energy wasted the voltage must be as .... as possible. Transformers are used to .... the voltage produced in power stations before transmission. This reduces the size of the .... so less energy is lost.

2. Describe the differences in structure between a step-down and a step-up transformer. [4]

   ....

   ....

   ....

   ....

3. The diagram shows an isolating transformer.

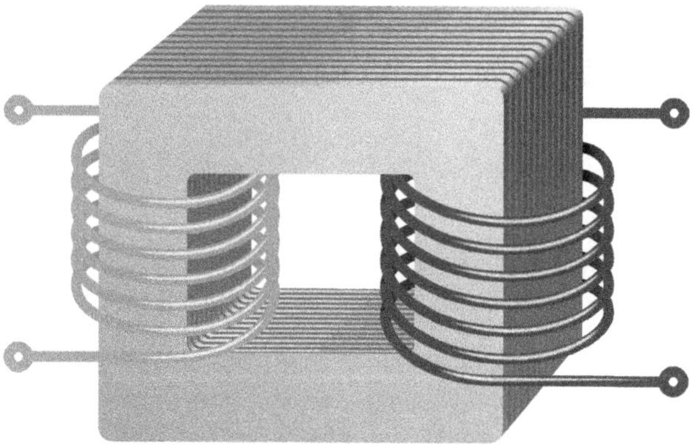

   Describe how isolating transformers reduce the risk of electrocution. [3]

   ....

   ....

   ....

   ....

[Total: .... / 9]

Higher Tier

4. A transformer has an input voltage, V_p, of 400 V. The primary coil has 1000 turns and the secondary coil has 50 turns.

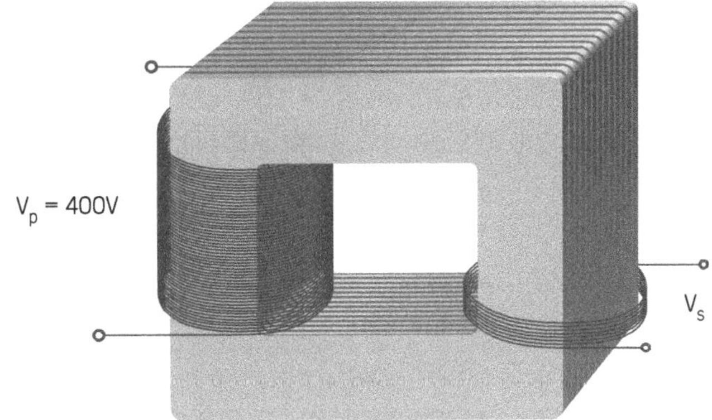

(a) Is this a step-up or a step-down transformer? .... [1]

(b) Calculate the output voltage across the secondary coil. [3]

....

....

....

5. The power loss in electrical wires is proportional to the current **squared**. An electrical wire has a resistance of 100 Ω.

(a) Calculate the loss when a current of 4 A flows through it. [3]

....

....

....

(b) The power loss is 400. What current was flowing through the wire? [2]

....

....

[Total: .... / 9]

1. The diagram shows a current–voltage graph.

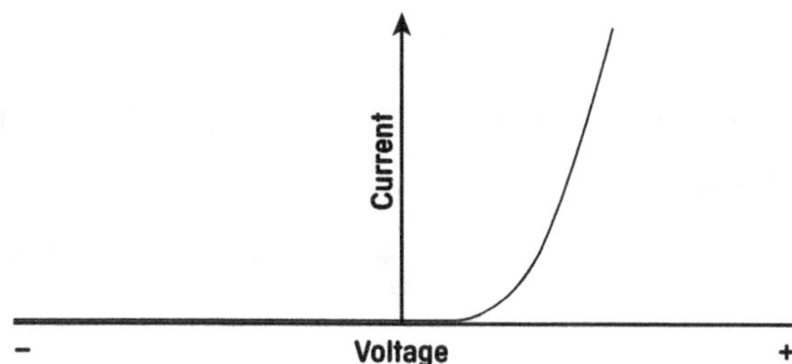

(a) Which component has current-voltage characteristics like those shown on the graph? [1]

....

(b) Draw the circuit symbol for this component. [1]

(c) How does this graph explain how a diode works? [2]

....

....

2. The graph below shows an unrectified wave.

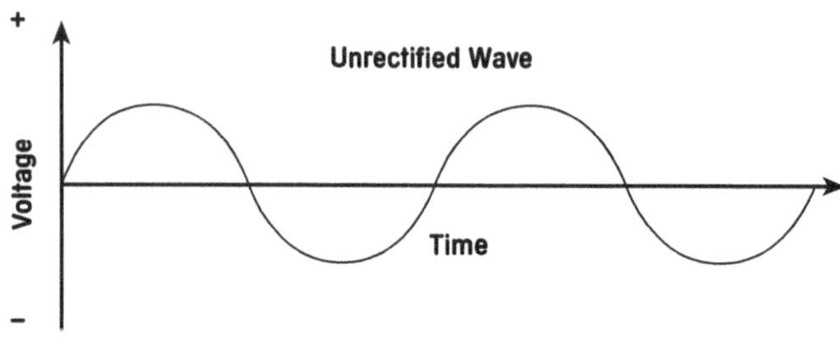

(a) Draw the voltage–time graph to show half-wave rectification of this wave. [2]

(b) What component produces half-wave rectification? [1]

....

3. The diagram shows a bridge circuit.

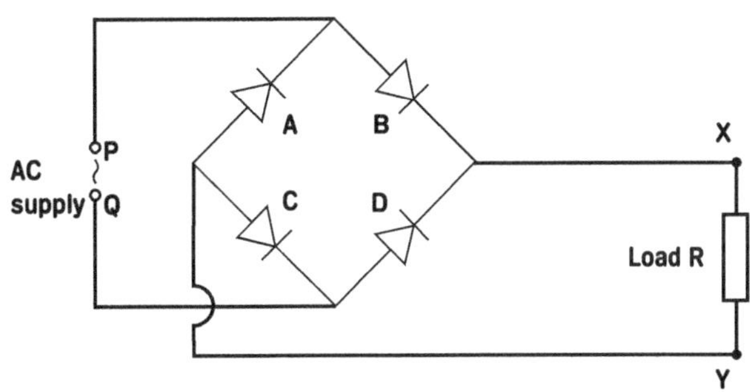

(a) During the positive half of the cycle the current passes diode **B**. Which other diode does the current pass through during this half of the cycle? .... [1]

(b) What type of rectification does a bridge circuit produce? [1]

....

4. A circuit has a capacitor and a voltmeter. The capacitor was uncharged when the circuit was first assembled.

(a) A battery is added to the circuit, as shown in the diagram below. What will happen to the capacitor when the battery is connected to the circuit? What would the voltmeter show? [2]

....

....

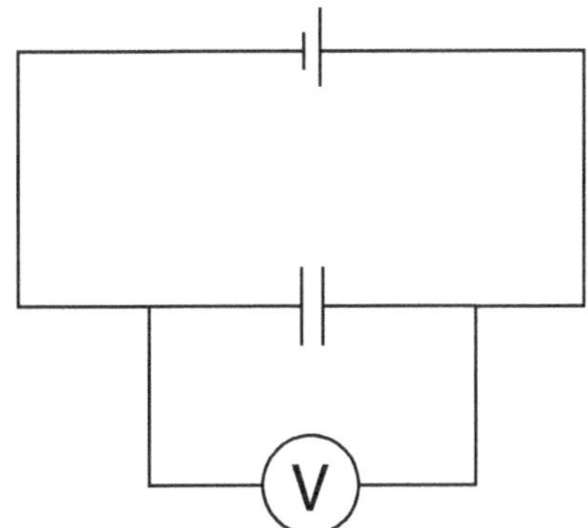

(b) The battery is then removed from the circuit. What would happen to the flow of current in the circuit over time and what would the voltmeter show? [3]

....

....

....

[Total: .... / 14]

Higher Tier

5. Explain the action of a capacitor in a simple smoothing circuit. [3]

....

....

....

[Total: .... / 3]

Notes

The Periodic Table

Key

relative atomic mass **atomic symbol** name atomic (proton) number

1	2											3	4	5	6	7	0
							1 **H** hydrogen 1										4 **He** helium 2
7 **Li** lithium 3	9 **Be** beryllium 4											11 **B** boron 5	12 **C** carbon 6	14 **N** nitrogen 7	16 **O** oxygen 8	19 **F** fluorine 9	20 **Ne** neon 10
23 **Na** sodium 11	24 **Mg** magnesium 12											27 **Al** aluminium 13	28 **Si** silicon 14	31 **P** phosphorus 15	32 **S** sulfur 16	35.5 **Cl** chlorine 17	40 **Ar** argon 18
39 **K** potassium 19	40 **Ca** calcium 20	45 **Sc** scandium 21	48 **Ti** titanium 22	51 **V** vanadium 23	52 **Cr** chromium 24	55 **Mn** manganese 25	56 **Fe** iron 26	59 **Co** cobalt 27	59 **Ni** nickel 28	63.5 **Cu** copper 29	65 **Zn** zinc 30	70 **Ga** gallium 31	73 **Ge** germanium 32	75 **As** arsenic 33	79 **Se** selenium 34	80 **Br** bromine 35	84 **Kr** krypton 36
85 **Rb** rubidium 37	88 **Sr** strontium 38	89 **Y** yttrium 39	91 **Zr** zirconium 40	93 **Nb** niobium 41	96 **Mo** molybdenum 42	[98] **Tc** technetium 43	101 **Ru** ruthenium 44	103 **Rh** rhodium 45	106 **Pd** palladium 46	108 **Ag** silver 47	112 **Cd** cadmium 48	115 **In** indium 49	119 **Sn** tin 50	122 **Sb** antimony 51	128 **Te** tellurium 52	127 **I** iodine 53	131 **Xe** xenon 54
133 **Cs** caesium 55	137 **Ba** barium 56	139 **La*** lanthanum 57	178 **Hf** hafnium 72	181 **Ta** tantalum 73	184 **W** tungsten 74	186 **Re** rhenium 75	190 **Os** osmium 76	192 **Ir** iridium 77	195 **Pt** platinum 78	197 **Au** gold 79	201 **Hg** mercury 80	204 **Tl** thallium 81	207 **Pb** lead 82	209 **Bi** bismuth 83	[209] **Po** polonium 84	[210] **At** astatine 85	[222] **Rn** radon 86
[223] **Fr** francium 87	[226] **Ra** radium 88	[227] **Ac*** actinium 89	[261] **Rf** rutherfordium 104	[262] **Db** dubnium 105	[266] **Sg** seaborgium 106	[264] **Bh** bohrium 107	[277] **Hs** hassium 108	[268] **Mt** meitnerium 109	[271] **Ds** darmstadtium 110	[272] **Rg** roentgenium 111	Elements with atomic numbers 112–116 have been reported but not fully authenticated						

*The lanthanoids (atomic numbers 58–71) and the actinoids (atomic numbers 90–103) have been omitted.

The relative atomic masses of copper and chlorine have not been rounded to the nearest whole number.

Equations

Average speed $= \frac{\text{Distance}}{\text{Time}}$

Acceleration $= \frac{\text{Change in speed}}{\text{Time taken}}$

Force = Mass × Acceleration

Work done = Force × Distance

Distance = Average speed × Time

Power $= \frac{\text{Work done}}{\text{Time}}$

$KE = \frac{1}{2} mv^2$

Weight = Mass × Gravitational field strength

Force $= \frac{\text{Change in momentum}}{\text{Time}}$

Refractive index $= \frac{\text{Speed of light in vacuum}}{\text{Speed of light in medium}}$

Power = Force × Speed

Power = Voltage × Current

Energy = Mass × Specific heat capacity × Temperature change

Energy = Mass × Specific latent heat

Efficiency $= \frac{\text{Useful energy output } (\times 100\%)}{\text{Total energy input}}$

Resistance $= \frac{\text{Voltage}}{\text{Current}}$

$v = u + at$

$s = \frac{(u + v)}{2} \times t$

Momentum = Mass × Velocity

Magnification $= \frac{\text{Image size}}{\text{Object size}}$

$V_p I_p = V_s I_s$

$v^2 = u^2 + 2as$

$s = ut + \frac{1}{2} at^2$

$I_e = I_b + I_C$

Wave speed = Frequency x Wavelength

Energy supplied = Power x Time

$GPE = mgh$

$mgh = \frac{1}{2} mv^2$

$m_1u_1 + m_2u_2 = (m_1 + m_2)v$

Power loss = $(\text{Current})^2$ × Resistance

$\frac{\text{Voltage across primary coil}}{\text{Voltage across secondary coil}} = \frac{\text{Number of primary turns}}{\text{Number of secondary turns}}$

GCSE OCR Gateway Physics B Workbook Answers

Answering Quality of Written Communication Questions

A number of the questions in your examinations will include an assessment of the quality of your written communication (QWC). These questions are worth a maximum of 6 marks and are indicated by a pencil icon (✐).

Your answers to these questions will be marked according to...
- the level of your understanding of the relevant science
- how well you structure your answer
- the style of your writing, including the quality of your punctuation, grammar and spelling.

QWC questions will be marked using a 'Levels of Response' mark scheme. The examiner will decide whether your answer is in the top level, middle level or bottom level. The expected quality of written communication is different in the three levels and it will always be considered at the same time as looking at the scientific information in your answer:
- To achieve Level 3 (which is the top level and is worth 5–6 marks), your answer should contain relevant science, and be organised and presented in a structured and coherent manner. You should use scientific terms appropriately and your spelling, punctuation and grammar should have very few errors.
- For Level 2 (worth 3–4 marks), there may be more errors in your spelling, punctuation and grammar, and your answer will miss some of the things expected at Level 3.
- For Level 1 (worth 1–2 marks), your answer may be written using simplistic language. You will have included some relevant science, but the quality of your written communication may have limited how well the examiner can understand your answer. This could be due to lots of errors in spelling, punctuation and grammar, misuse of scientific terms or a poor structure.
- An answer given Level 0 may contain insufficient or irrelevant science, and will not be given any marks.

You will be awarded the higher or lower mark within a particular level depending on the quality of the science and the quality of the written communication in your answer.

Even if the quality of your written communication is perfect, the level you are awarded will be limited if you show little understanding of the relevant science, and you will be given Level 0 if you show no relevant scientific understanding at all.

To help you understand the criteria above, three specimen answers are provided to the first QWC question in this workbook. The first is a model answer worth 6 marks, the second answer would be worth 4 marks and the third answer worth 2 marks. The three exemplar answers are differentiated by their scientific content and use of scientific terminology. Model answers worth 6 marks are provided to all other QWC questions to help you aspire to the best possible marks.

P1 Energy for the Home

Pages 3–5

1. joule/J
2. **(a)** Thermogram
   (b) Windows
3. **(a)** Helium; Highest specific heat capacity (SHC)
   (b) Energy = mass × SHC × change in temperature; Energy = 2 × 380 × 10; Energy = 7600 J. **(1 mark for calculation, 1 mark for correct answer.)**
4. The temperature stays the same; So the chocolate must be melting.
5. **(a)** Material **and** State change (solid to liquid/liquid to gas).
   (b) Energy = mass × specific latent heat; Energy = 20 × 330; Energy = 6600 J. **(1 mark for calculation, 1 mark for correct answer.)**
6. mass = energy ÷ (SHC × temperature change); mass = 660000 ÷ (440 × 15) = 100 kg (**1 mark for calculation, 1 mark for correct answer**).
7. Temperature is a measurement of the average kinetic energy of particles; Heat is a measurement of energy on an absolute scale.
8. SHC = energy ÷ (mass × change in temperature); SHC = 7500 ÷ (0.5 × 6); SHC = 2500 J/kg°C. **(1 mark for calculation, 1 mark for correct answer.)**
9. **(a)** Break bonds; change state **(1 mark each)** but break intermolecular bonds **would get 2 marks**.
   (b) energy = mass × specific latent heat, rearrange to find SLH, SLH = energy ÷ mass, 66.8 ÷ 0.2 = 334 kJ/kg **(1 mark for calculation, 1 mark for correct answer)**.

Pages 6–8

1. **(a)** Infrared radiation is reflected by a dull or shiny surface.
   (b) Infrared radiation is absorbed by a rough surface.
2. **(a)** Cavity wall insulation
   (b) Loft insulation; Will take 6 months to save the original cost.
3. **(a)** 30,000 – 12,000 = 18,000 J
   (b) Efficiency = useful energy output ÷ total energy input; Efficiency = 18,000 ÷ 30,000; Efficiency = 0.6. (To put this into a percentage, multiply the answer by 100 = 60%). **(1 mark for calculation, 1 mark for correct answer.)**
   (c) It is less efficient than Jane's hairdryer (50% or 0.5 is lower than 0.6).
4. **(a)** Light bulb A: (11 ÷ 55) x 100 = 20%
   Light bulb B: (5 ÷ 50) × 100 = 10%
   Light bulb C: (18 ÷ 60) x 100 = 30%
   (2 marks for calculations)
   Light bulb C is most efficient. **(1 mark)**
   (b) 30% – 10% = 20%
5. **This is a model answer which would score the full 6 marks:** Heat loss by conduction through windows can be reduced by installing double-glazing. The gap between the two panes of glass is filled with air, which is a good insulator. The narrow gap also reduces heat loss by convection. Heat loss through walls can be reduced by using cavity wall insulation. Insulation material is placed inside a gap in the wall, which reduces the heat loss. Fibreglass loft insulation works in a similar way to cavity wall insulation, trapping layers of air between the fibres and reducing heat loss by conduction and convection. Shiny foil can be placed on walls behind radiators to reflect heat energy back into the room.
   This answer would score 4 marks: Double-glazing prevents heat loss by conduction through windows. Cavity wall insulation helps prevent heat loss through the walls because the cavity is filled with insulation material. Loft insulation traps air and reduces heat loss by conduction and convection.
   This answer would score 2 marks: Double-glazing stops too much heat escaping out of the windows. Cavity wall insulation uses insulating material to stop too much heat escaping through the walls.

6. Conduction: Transfer of KE; Between particles (including free electrons).
Convection: Expansion when a liquid or gas is heated; Causes a decrease in density which results in fluid flow.
Radiation: Infrared radiation is an electromagnetic wave; Needs no medium.
7. Useful energy = 64 J a second, wastes 16 J a second **(1 mark)**, 16 × 60 × 60 **(1 mark)** = 57.6 kJ wasted in an hour **(1 mark)**.

Page 9

1. **(a)** **A:** peak; **B:** trough; **C:** amplitude; **D:** wavelength.
(b) Frequency = 120 waves ÷ 30 seconds; Frequency = 4 Hz. **(1 mark for calculation, 1 mark for correct answer.)**
(c) Wave speed = frequency × wavelength, 4 × 50 = 200 m/s **(1 mark for calculation, 1 mark for correct answer)**.
2. Microwave; Infrared; Visible light; Ultraviolet; X-rays
3. Wavelength = speed ÷ frequency; Wavelength = 330 ÷ 194; Wavelength = 1.70 m. **(1 mark for calculation, 1 mark for correct answer.)**

Page 10

1. **Any two from**: Light turned on and off; Produce short and long pulses; Different combinations of pulses code for different letters.
2. **(a)** **(i)** **B**
(ii) **C**
3. Pits store digital information; Laser/light shone onto surface of CD; Light reflected from pits; Reflected pulses converted into (electrical) signal.

Pages 11–12

1. **(a)** Water **and** fat
(b) About 1 cm
(c) Infrared
2. **(a)** Mobile phone masts use microwave radiation; Which may be harmful to humans; Although there is no proven link between health problems and mobile phone masts.
(b) Check the results; Of published scientific studies into the health effects of mobile phone masts.
3. **(a)** Large obstacles in the way (i.e. trees/mountains); Poor weather conditions; Curvature of Earth; Interference between signals.
(b) Reduce distance between transmitters; Put masts on hills/high buildings.
4. Most of the microwaves are absorbed by water and fat particles; In the outer layers of the food; Increasing their kinetic energy; Some microwaves may pass to the middle of the food and be absorbed; Or kinetic energy is transferred from the outer layers to the centre by conduction or convection.

Pages 13–14

1. **Any two from**: Remote controls; Automatic door sensors; Burglar alarms; Security lights; Computer data links; Cooking
2. **(a)** **(i)** **A** is a digital signal, **B** is an analogue signal. **(1 mark)**.
(ii) Digital signal is on or off; Analogue signal is continuous.
(b) Digital; Digital signals have only two states (on or off) so are therefore less susceptible to noise.
(c) 0010010101
3. **(a)** Two or more signals sent down an optical fibre at the same time/simultaneously.
(b) **Any two from**: More data sent simultaneously; Multiple users; Faster transmission.

Pages 15–16

1. **(a)** **Any two from**: Radios; Mobile phones; Laptops (wireless internet); Remote controls.
(b) **Any two from**: Signals available 24 hours a day; No wiring needed so available in remote places; Portable.
2. Anyone with a DAB radio can receive digital radio. [✗]
DAB radios can receive digital broadcasts and old analogue signals. [✓]
There are far more digital stations available than analogue stations. [✓]
Digital signals sometimes contain more interference than analogue signals. [✗]
3. **(a)** Both radio stations are transmitting on a similar frequency so their analogue signals; Are interfering with each other, leading to poor reception.
(b) **Any one of:** One station changes its frequency; Either station switches to digital transmission.
4. Long wave signals; reflected by the ionosphere/back to Earth (around the curve of the Earth).
5. **(a)** Longwave signals diffracted by hill; Changes direction and reaches radio antenna; Shorter wavelength signal not diffracted.
(b) Diffraction causes wave to spread out which results in some signal loss; Interference caused by signal reflected by hills.

Pages 17–18

1. **(a)** Seismometer
(b) **Any three from: P-waves** are longitudinal; Faster than s-waves; Can travel through solids and liquids; **S-waves** are transverse; Slower than p-waves so arrive after p-waves at seismometer; Can only travel through solids.
2. **(a)** Cause sunburn/skin cancer
(b) Sun Protection Factor
(c) Safe time = time without sun block × SPF; 3 × 20 = 60 minutes; 10 × 20 = 200 minutes; 25 × 20 = 500 minutes. **(1 mark for each correct answer.)**
3. Scientists repeated their measurements with new equipment; Different scientists also repeated the same measurements and found the same results.
4. S-waves are transverse so; Cannot pass through liquid; Shadow zone shows; Outer core is liquid.
5. Hole in ozone layer shows human activities (use of CFCs) reducing ozone levels; Ozone absorbs UV radiation; Continued use of CFCs could lead to holes developing over populated areas; Hugely increasing risk of skin cancer due to increased exposure to UV radiation; International agreement on banning CFCs to prevent this.

P2 Living for the Future (Energy Resources)

Pages 19–20

1. **(a)** Light
(b) Increase surface area; More light captured; Higher electrical energy output.
(c) **Any three from:** Radiation absorbed and transferred to heat energy; Produces convection currents to drive wind turbines; Passive solar heating using glass; Reflected to a focus by a mirror.
2. **(a)** **Any three from:** Low maintenance; No need for fuel; Long life; No polluting waste; Reduce energy costs.
(b) Best position for the panel to absorb maximum sunlight; No power at night or in bad weather so need to have other sources of electricity.
3. **Advantages:** Renewable; No polluting waste; Doesn't contribute to the greenhouse effect and global warming.
Disadvantages: Dependent on wind speed so not a constant supply – must have a back-up power-generating source; Visual pollution.
4. **(a)** Increased distance from light source decreases light intensity; So decreases output.
(b) Surface exposed could also affect the output.
(c) **Any six from:** Passive solar heating: Glass is transparent so Sun's radiation passes through it; Heated surfaces emit infrared radiation of longer wavelength; Glass reflects this longer wavelength infrared so inside of building warms up. Photocell: Energy from Sun absorbed by photocell; Electrons are knocked loose from the silicon atoms in the crystal; Electrons flow freely.

Pages 21–22

1. **(a)** Moving a magnet; Within a coil of wire.
(b) **Any two from:** Move the magnet more quickly; Use a stronger magnet; Increase the number of turns on the coil of wire.
2. **(a)** Alternating current; Electric charge is reversing direction.
(b) A generator
3. Heats water – steam produced – turns turbine – turns generator. **(All four correct, 2 marks; two or three correct, 1 mark.)**
4. Efficiency = useful output ÷ total input (× 100); Efficiency = 3 MW ÷ 12 MW; Efficiency = 0.25 (or × 100 = 25%). **(1 mark for calculation, 1 mark for correct answer.)**

5. 125,000 J useful energy out = 25% of total in; Total input = 4 × 125,000 J; Input = 500,000 J. **(2 marks for calculation, 1 mark for correct answer.)**

Pages 23–24

1. **Supports:** Atmospheric carbon dioxide levels are rising due to human activity; Increased carbon dioxide levels leads to global warming from the greenhouse effect.
   Refutes: Solar radiation reaching the Earth has increased at the same time as the temperature has increased; Humans don't affect solar radiation.
2. Cutting down large numbers of trees; Less carbon dioxide taken from the atmosphere by the photosynthesis of trees; Increased carbon dioxide levels lead to global warming.
3. **Any two from:** Methane; Carbon dioxide; Water vapour; Nitrous oxide; Ozone.
4. **This is a model answer, which demonstrates QWC, and would therefore score the full 6 marks:** The greenhouse effect is a scientific process that can be observed and experimented on so it's easy for scientists to agree on it. Scientists can't agree on human activity affecting global warming as it's a very complex issue that is not fully understood. The number of interacting factors that affect global warming mean it's very difficult to isolate one factor (human activity) and categorically state that it is having an effect. While there is a large amount of evidence that supports the idea that human activity is leading to global warming there is other evidence that refutes this idea.
5. **(a)** The global temperature increased steadily up until 1940; It then decreased at a similar rate for the next 20 years; It then rose again up to 2000.
   (b) 0.4°C (**Accept** 0.3°C **or** 0.5°C)

Pages 25–26

1.

Type of fuel used in power stations	Example of fuel	Advantage	Disadvantage
Fossil fuels	Coal/oil/gas	Reliable	Will run out/ produces carbon dioxide that leads to global warming.
Renewable biomass	Manure/ straw/ wood	Renewable so won't run out.	Still produce some carbon dioxide that leads to global warming.
Nuclear fuels	Uranium/ plutonium	Reliable, produces very little carbon dioxide.	Fuel and waste is very dangerous, high cost in disposing of waste.

(2 marks for each correct row.)

2. Power = current × voltage (P = I × V); P = 10 × 230; P = 2300 W. **(1 mark for calculation, 1 mark for correct answer.)**
3. **(a)** Power rating of device ÷ time appliance used for in hours.
   (b) Cost = units × cost per unit; Cost = 4 × 10; Cost = 40 pence. **(1 mark for calculation, 1 mark for correct answer.)**
4. **(a)**

Appliance	Power rating in kilowatts	Average time used each week (hours)	Total energy used in a week (kWh)
Hairdryer	2.3	1	**2.3**
Dishwasher	1.2	**2.5**	3
Washing machine	**1.7**	4	6.8
TV	0.12	**10**	1.2

   (b) Power = 2.3 KW = 2300 W
   Power = current × voltage
   Current = power ÷ voltage
   Current = 2300 ÷ 230
   Current = 10 A
   (1 mark for calculation, 1 mark for correct answer.)
5. **Advantages:** Less demand at night; Cheaper; Avoids wasting electricity.
   Disadvantages: Appliances such as washing machines noisy to run at night.
6. Heating large amounts of cables in the national grid would lead to a large amount of energy wasted; The national grid solves this problem by increasing the voltage; This decreases the current; So lowers the temperature of wires, so less energy wasted as heat.

Pages 27–28

1. Protective clothing; Use shielding (i.e. lead apron); Keep distance between her and substance as great as possible; Keep exposure time to a minimum.
2. **Alpha:** Smoke detectors.
   Beta: Any one from: Tracers; Paper thickness gauges.
   Gamma: Any one from: Treating cancer; Tracers; Non-destructive testing; Sterilising equipment.
3. **(a)** A few mm aluminium
   (b) Alpha
   (c) Gamma
4. Gamma
5. **Positive ions:** When electrons are lost from atoms.
   Negative ions: When electrons are gained by atoms.
6. **(a)** Plutonium
   (b) **Low level**: sealed in canisters and stored/buried in landfill. **High level**: encased in glass and stored underground.
7. **(a)** **1**= Beta; stopped by thin sheet of aluminium; **2** = Gamma; only stopped by lead; **3** = Alpha; stopped by paper.
   (b) Carry out repeat experiments.

Pages 29–30

1. Planet, star, galaxy. **(1 mark for each correct label and 1 mark for the correct order.)**
2. A light year is the distance light travels in one year.
3. **(a)** An unmanned probe; Temperature and atmospheric conditions; Would make it very hostile and dangerous for humans.
   (b) Due to the large distance between Earth and Venus; Data would be sent back from planet; As the Moon is much closer to Earth; Actual samples could be sent back.
4.

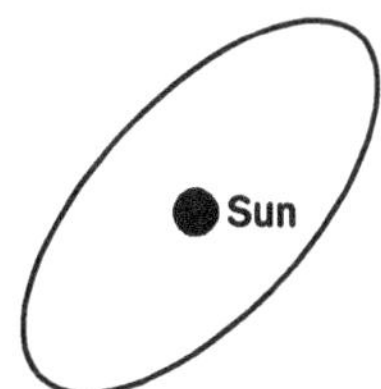

5. The gravitational attraction; Of the Sun.
6. **Any two from:** Kilometres used for distances within the solar system; Light years used for distances between solar systems; Across our galaxy; Between galaxies.

Pages 31–32

1. **(a)** Asteroid is made of rocks; **Any one of:** Comet is made from and ice and dust; Has a tail formed from trail of debris.
   (b) Craters; Sudden change in the fossil record; Unusual chemical elements in the rocks.
   (c) Iron cores merge to form the core of the Earth; Less dense material orbits and forms the Moon.
2. **(a)** Near Earth Object
   (b) They may be on a collision course with Earth.
3. Jupiter's large gravity; Pulls them apart.
4. The further from the Sun, the slower the comet; As the comet gets closer to the Sun its speed increases; Due to the gravitational attraction of the Sun.
5. Survey the skies using telescopes; Monitor object's progress (trajectory); Could deflect object.

6. Samples from the Moon's surface indicate it was once molten; Moon's small iron core; Similar collisions occur in other star systems.

Pages 33–34

1. **(a)** **Any two from:** Formed billions of years ago; Huge explosion; Universe still expanding.
   (b) All galaxies moving away from us (universe expanding); Distant galaxies are moving away from us faster; Detect microwave radiation from all parts of the universe.
2. **(a)** Gravity
   (b) Hydrogen
3. Star shrinks rapidly, then explodes to form a supernova; The remnants of the star form a neutron star; A black hole is left behind.
4. **(a)** **Any two from:** The church was very influential and believed that the Earth was the centre of the universe; The Sun seems to move across the sky; Technology had not progressed sufficiently to make good observations.
   (b) **Any two from:** Venus did not move smoothly across sky; Retrograde motion; Orbit of planets not solely due to rotation of Earth.
5. **(a)** Light from galaxies moving away from us; Is further towards the red end of the visible spectrum than expected.
   (b) Light from distant galaxies; Much more red-shifted.
   (c) The fate of the universe

P3 Forces for Transport

Pages 35–36

1. Distance it travelled **and** time it took to travel that distance.
2. Speed = distance ÷ time; Speed = 3000 ÷ 150; Speed = 20 metres per second. **(1 mark for the correct answer and 1 mark for the correct unit.)**
3. **0–6 seconds** – car is stationary.
   6–7 seconds – car is moving forward.
   7–9 seconds – car is continuing to move forward but at a lower speed.
4. The speed may speed up/slow down.
5. **(a)** S = d ÷ t; S = 6000 ÷ 300; S = 20 metres per second. **(1 mark for calculation, 1 mark for correct answer.)**
   (b) t = d ÷ s; t = 50000 ÷ 20; t = 2500 s (41 minutes 40 seconds) **(1 mark for calculation, 1 mark for correct answer.)**
6. **Object 2**: s = d ÷ t; s = 40 ÷ 25
   s = 1.6 m/s
   Object 3: s = d ÷ t; s = 60 ÷ 25
   s = 2.4 m/s

Pages 37–38

1. Change in speed **and** time taken.
2. **(a)** Acceleration = change in speed ÷ time taken; Acceleration = 5 – 0 ÷ 5; Acceleration = 1 m/s^2 **(1 mark for calculation, 1 mark for correct answer, 1 mark for unit.)**
   (b) Cyclist **B** has higher acceleration. It has greater speed after the same time.
3. The second cyclist is travelling in the opposite direction to the first cyclist.
4. **(a)** 6 seconds
   (b) A constant; Negative gradient.
5. Distance travelled
6. Time taken = change in speed ÷ acceleration; Time = (4 – 0) ÷ 10; Time = 0.4 seconds. **(2 marks for correct calculation including rearranging equation, 1 mark for correct answer.)**
7. **(a)** Acceleration = change in speed ÷ time taken; Acceleration = (15 – 0) ÷ 3; Acceleration = 5 m/s^2. **(1 mark for calculation, 1 mark for correct answer, 1 mark for unit.)**
   (b) Distance 0–3 seconds = 0.5 × 3 × 15 = 22.5 m **(1 mark)**
   Distance 3–6 seconds = 3 × 15 = 45 m **(1 mark)**
   45 + 22.5 = 67.5 m **(1 mark)**

Pages 39–40

1. F = m × a; F = 1.5 × 2; F = 3 N **(1 mark for calculation, 1 mark for correct answer.)**
2. Lorry **B** accelerates fastest; Smaller mass accelerates greatest when same force is used.
3. **A:** Thinking distance
   B: Braking distance
   C: Stopping distance
4. **(a)** **Braking distance:** Ice on the roads; Wear on brakes **(1 mark)**
   Thinking distance: Tiredness; Alcohol intake **(1 mark)**
   (b) Tiredness is subjective so is therefore very difficult to measure.
5. **(a)** Acceleration = (6 – 0) ÷ 2; Acceleration = 3 m/s^2. **(1 mark for calculation, 1 mark for correct answer.)**
   (b) F = m × a; F = 900 × 3; F = 2700 N. **(1 mark for calculation, 1 mark for correct answer.)**
   (c) A = F ÷ m; A = 2700 N ÷ 600 kg; A = 4.5 m/s^2. **(1 mark for calculation, 1 mark for correct answer.)**
6. Icy roads reduce friction between the tyres and the road; Therefore the brakes may stop the wheels of the car spinning but not stop them moving (the car slides) so the car doesn't decelerate as quickly; Increasing the braking distance.
7. Thinking distance increases linearly with speed (double speed, double thinking distance); Braking distance increases as a squared relationship (double speed, increase braking distance by a factor of four).

Pages 41–42

1. Joules
2. Weight **and** height **(both needed for 1 mark)**.
3. **(a)** Work done = force (weight) × distance; Work done = 24 × 0.5; Work done = 12 J. **(1 mark for calculation, 1 mark for correct answer.)**
   (b) Energy transferred = work done so energy transferred = 12 J.
   (c) power = WD ÷ time taken
   power = 12 ÷ 3
   power = 4 W **(1 mark for calculation, 1 mark for correct answer.)**
   (d) mass × gfs = 10 x 10 = 100 N
   (e) Original work done = 12 J
   Distance = work done ÷ force
   12 ÷ 100 = 0.12 m **(1 mark for rearrangement and calculation, 1 mark for the answer.)**
4. **(a)** Car **A** **(1 mark)**; **Any two from:** Best fuel consumption value – most km for each litre of fuel; Less fuel used per journey than other cars; Lowest CO_2 emissions per km.
   (b) Yes, as it uses the least fuel per km; Therefore cost of petrol per km will be the lowest.
5. **(a)** mass = weight ÷ gravitational field strength
   (3000 + 5000) ÷ 10 = 800 kg **(1 mark for rearrangement and calculation, 1 mark for the answer)**
   (b) work done = force × distance
   8000 × 6 = 48 kJ
   (c) power = work done ÷ time
   = 48000 ÷ 10 = 4800 W

Pages 43–44

1. Speed **and** mass **(both needed for 1 mark)**.
2. KE = 0.5 × m × v^2
   Car = 0.5 × 1000 × 30^2
   = 450 kJ
   Lorry = 0.5 × 5000 × 30^2
   = 2250 kJ
   (1 mark for two correct calculations and 2 marks for correct answers.)
3. **(a)** Use of deflectors on the cab roof.
   (b) Fuel consumption will increase – less km per litre.
4. Electric cars have to have their batteries recharged; The electricity from this comes from power stations that may cause pollution.
5. Petrol and diesel fuel comes from oil; Which is a non-renewable energy source and will run out; Solar and bio fuel are renewable energy sources so won't run out.

6. **This is a model answer, which demonstrates QWC, and would therefore score the full 6 marks:** Fuel consumption depends on a number of factors. If the driver accelerates, an increase in kinetic energy is required so more fuel is burnt. Therefore a driver who accelerates often and drives at high speed will have higher fuel consumption than a driver who does not. Smooth, good quality roads decrease friction between tyres and the road, and having tyres in good condition also achieves this. A decrease in friction reduces the force required to overcome it, therefore less energy is required so a factor that lowers friction also decreases fuel consumption.
7. mass = (KE ÷ v^2) × 2
   = (72000 ÷ 12^2) × 2 = 1000 kg
   (1 mark for rearrangement and calculation, 1 mark for the answer.)
8. His kinetic energy quadruples (multiplies by four); Kinetic energy is proportional to 'velocity squared'.

Pages 45–46

1. **Any two from:** Anti-lock braking system (ABS); Traction control; Electric windows; Paddle shift controls.
2. Change shape during collision; Absorb energy; Protect occupants/reduce injury.
3. **(a)** Total mass of driver and occupants = 1000 + 200 kg = 1200 kg
   (b) Momentum = mass × velocity; Momentum = 1200 × 20; Momentum = 24000 kgm/s. **(1 mark for calculation, 1 mark for correct answer.)**
   (c) force = change in momentum ÷ time
   = 24000 ÷ 0.5 = 48000 N
   (1 mark for calculation, 1 mark for correct answer.)
4. Make it possible to keep control of steering in hazardous conditions (e.g. skid); Brakes automatically pump on and off to avoid skidding; Sometimes reduce braking distance.
5. **Any four from:** A moving car has kinetic energy; During collision crumple zones change shape/absorb or dissipate energy; Crumple zones increase time taken to decelerate; Force is proportional to rate of deceleration (F = m × a); So if acceleration decreases, force felt by occupants decreases.

Pages 47–48

1. **(a)** **X** is air resistance; **Y** is weight (**accept** gravity).
   (b) Gravity
   (c) Increases
   (d) **(i)** Terminal speed/terminal velocity
   (ii) **Any two from:** Forces **X** and **Y** are the same size; Air resistance and weight are the same size; Weight and air resistance are balanced.
2. **(a)** **between** 4 **and** 5.
   (b) Parachute larger surface area than body so air resistance is greater; Air resistance is greater than weight so skydiver slows down/decelerates; Air resistance equals weight at lower speed.
3. **(a)** Weight = mass × acceleration due to gravity; Weight = 70 × 10; Weight = 700 N.
   (b) 700 N; At terminal velocity air resistance equals weight. **(1 mark for value, 1 mark for explanation.)**
   (c) Location on the Earth's surface; Height above or below the Earth's surface (e.g. down a mineshaft).

Pages 49–51

1. **(a)** The skydiver has gravitational potential energy.
   (b) The position of the skydiver in the Earth's gravitational field/ height of the skydiver; The mass of the skydiver.
   (c) It will decrease
2. **(a)** The hammer; Greatest mass.
   (b) GPE = mgh
   = 2 × 1.6 × 1.2
   = 3.84 J **(1 mark for calculation, 1 mark for correct answer.)**
3. **(a)** Gravitational potential energy
   (b) Changes into kinetic energy
   (c) At 3
   (d) The kinetic energy would double.
4. mass = GPE ÷ h ÷ g
   = 490 ÷ 3.5 ÷ 10 = 14 kg
   (1 mark for calculation, 1 mark for correct answer)
5. As he initially falls GPE is converted to KPE; When terminal velocity is reached KPE is no longer increasing; GPE is transferred to increased internal or thermal energy of the surrounding air particles; Through friction.
6. Double the speed; Speed is squared in the kinetic energy formula.
7. 70 × 200 × 10 = 140,000 J **(1 mark for calculation, 1 mark for correct answer)**

P4 Radiation for Life

Pages 52–53

1. insulating, electrons, attract.
2. Charge transferred to him from carpet/charge builds up on him; The metal tap is a conductor; Negatively-charged electrons transferred from Joe to tap; Down to earth.
3. **Nuisance – any two from**: TV screens or computer screens attract dirt; Clothes stick to skin; Static shock; Damage to sensitive electrical equipment in factory.
   Dangerous – any two from: Spark can ignite flammable gases or liquids; Spark can cause explosion in places with high oxygen concentration; Lightning can kill.
4. All materials become charged when rubbed with a cloth. [✗]
   Friction causes electrons to move from one material to another. [✓]
   A positively charged Perspex rod will repel a negatively charged piece of paper. [✗]
   When hair is combed the hair may stick up because each hair has the same charge. [✓]
5. **(a)** Electrons are transferred from the rod to the cloth.
   (b) A negative charge; Due to having an excess of electrons.
6. An atom or molecule with excess electrons/has gained electrons **(1 mark)**; So has a negative charge **(1 mark)**.
7. **(a)** Friction between fuel and nozzle; Charge builds up so plane becomes charged; Could cause spark; Spark could ignite fuel vapour.
   (b) By earthing

Pages 54–55

1. **Any two from**: Photocopiers; Laser printers; Defibrillators; Spray painting; Smoke precipitators.
2. **Any four from:** Negatively charged paint; Paint droplets repel each other; So fine spray; Car positively charged; Paint attracted.
3. **(a)** It contracts
   (b) Stand well back; Mary must not be touching patient or metal bed; Wear rubber-soled shoes/so insulated from ground/ Mary will be route to earth.
4. Smoke passes through positively-charged grid so loses electrons; Smoke is now positively-charged due to loss of electrons; Positively-charged smoke particles are attracted to negatively-charged plate.

Pages 56–57

1. **(a)** The current decreases as the resistance increases.
   (b) Ohm/Ω
   (c) As the voltage increases the current increases.
   (d) The current will decrease.
2. Resistance = voltage ÷ current; R = 200 V ÷ 0.5 A; R = 400Ω **(1 mark for calculation, 1 mark for correct answer)**
3. **A:** Neutral; Completes the circuit.
   B: Live; Carries the high voltage.
4. **Any two from:** Plastic (not metal) casing; Casing made from insulating material does not need earth wire; Cannot become live.
5. **Any four from:** Fault occurs; Current increases; Fuse wire melts; Circuit broken; No current can flow; Prevents flex over-heating and causing a fire and prevents any further damage to the appliance.
6. Power = current × voltage; Power = 11.3 A × 230 V; Power = 2599 W **(1 mark for calculation, 1 mark for correct answer.)**
7. **(a)** current = voltage ÷ resistance
   = 0.2 ÷ 0.05
   = 10 A
   (1 mark for substitution and calculation, 1 mark for the correct answer.)
   (b) **Any four from:** Casing becomes live; Short circuit; Current flows through earth wire (it has less resistance so current increases); Fuse melts; Circuit broken.

Pages 58–59

1. Ultrasound is a longitudinal wave.
2. (a) **Frequency:** Number of complete waves per unit of time (normally per second). **Wavelength:** Distance from one compression to another.
   (b) **Compression:** A region of higher pressure; **Rarefaction:** A region of lower pressure.
3. Frequency of ultrasound is higher than the upper limit of human hearing; Ear cannot detect very high frequencies.
4. Time to reach bottom = 0.5 ÷ 2 = 0.25 seconds
   Distance = speed × time
   1400 × 0.25 = 350
   Depth = 350 m
   (2 marks for calculation, 1 mark for correct answer)
5. Blood
6. Breaking down kidney stones.
7. **Any four from:** Ultrasound waves transmitted into body; Waves partially reflected at boundary between different materials in body; Time taken for reflected wave to be detected used to calculate; Depth of boundary; Image produced on screen.
8. See soft tissue; No damage to living cells.

Pages 60–61

1. Number of nuclear decays; Emitted per second.
2. (a) Alpha; Beta; Gamma.
   (b) Alpha particle is a helium nucleus; Beta particle is a fast-moving electron.
3. Time taken for the activity of a sample to decrease to half the starting value/time taken for half of the undecayed nuclei in a sample to decay.
4. (a) ${}^{232}_{90}\text{Th} \longrightarrow {}^{228}_{88}\text{Ra} \longrightarrow + {}^{4}_{2}\alpha$
   (1 mark for Ra, 1 mark for both atomic and mass number of Ra, 1 mark for including alpha particle ${}^{4}_{2}\alpha$.)
   (b) ${}^{228}_{89}\text{Ac} \longrightarrow {}^{228}_{90}\text{Th} + {}^{0}_{-1}\beta$
   (1 mark for Th, 1 mark for each of atomic and mass number of Th, 1 mark for including beta particle.)
5. (a) **(All points need to be plotted correctly to obtain 1 mark.)**

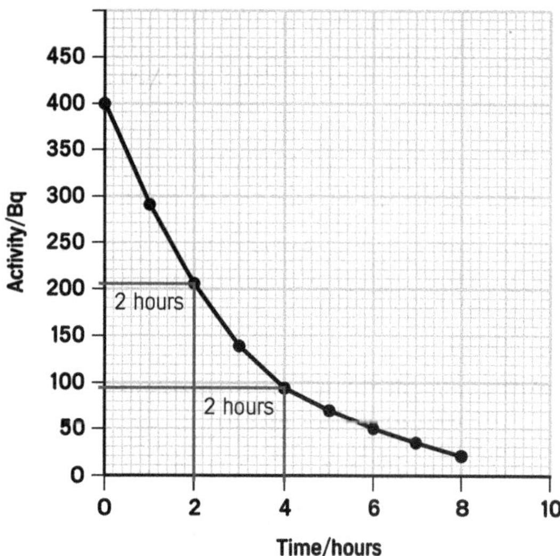

   (b) **(Curve of best fit needs to be drawn correctly for 1 mark.)**
   (c) Graph used to determine half-life, as shown above
   half-life = 2 hours

Page 62

1. Cosmic rays; Rocks and soil.
2. Radioactive material put into a pipe; Detector tracks progress of radioactive material; A reduction in radioactivity after a certain point indicates blockage.
3. **Any four from:** Alpha particle ionises air so; Electric current flows between electrodes; Smoke absorbs alpha particles so; No ionisation and reduced or no current; Alarm sounds.
4. Once-living organism ingests carbon-14; Whilst organism is alive carbon-14 which decays is replaced; After death the amount of carbon-14 decreases; Compare remaining amount of carbon-14 to similar living organism.

Page 63

1. **Similarities:** Both ionising electromagnetic waves; Similar wavelengths. **Difference:** Produced in different ways.
2. Technicium-99 has a short half-life/Technicium-99 decays quickly; Minimise damage to tissues inside body; Uranium-235 has a very long half-life/remains active for a very long period of time (millions of years).
3. **This is a model answer, which demonstrates QWC, and would therefore score the full 6 marks:** In radiotherapy gamma rays of a very specific dosage are focused on the tumour. As gamma rays are ionising radiation they kill the cells that make up the tumour. This then leads to the shrinkage or disappearance of the tumour. However there is a risk of the gamma rays damaging healthy tissue around the tumour, which could then lead to further health problems. The risks are minimised by ensuring the correct dosage of radiation is used and that it is very precisely targeted. The patient's doctors must weigh up the risks associated with the radiotherapy against the likelihood of it successfully treating the cancer.

Page 64

1. Nuclear fuel – atoms are split/releases heat energy; Heat used to turn water into steam; Steam turns turbine; Turbine turns generator, which produces electricity.
2. (a) The atom splits.
   (b) Requires very high temperatures; More energy is put in than is given out by the fusion.
3. If cold fusion could be carried out it would revolutionise energy production; Other scientists must be able to use the same method to produce similar results (to prove the experiment is repeatable); The data and method may help other scientists develop other areas of research.
4. (a) Uranium nucleus splits; More than one neutron is ejected; These extra neutrons cause further uranium nuclei to split.
   (b) Absorb some of the neutrons; Preventing them from hitting other atoms of uranium/causing further fissions.

P5 Space for Reflection

Pages 65–66

1. (a) An object that orbits another object/a planet in space.
   (b) Gravity/centripetal force
   (c) **Artificial satellites:** Objects which have been artificially placed into orbit by humans, e.g. communication satellites or space stations; **Natural satellites;** Objects which naturally orbit a planet or other body in space, e.g. the Moon.
2. Orbits the Earth once in 24 hours; Remains in a fixed position above the Earth's surface.
3. Low polar orbit more suitable as orbits Earth several times a day; Have very low orbit compared to geostationary; Much more suitable for getting pictures of target.
4. Centripetal force; Provided by gravity produced by Earth.
5. **This is a model answer, which demonstrates QWC, and would therefore score the full 6 marks:** The comet has an elliptical orbit around the Sun. This means the force of gravity from the Sun that is acting on it varies greatly during this orbit. When it's close to the Sun the gravitational force is strong so the comet is travelling at a relatively high speed. However as the comet moves away from the Sun the force of gravity decreases according to the inverse square law, e.g. a doubling of distance causes the force of gravity to reduce by a factor of four, whilst a trebling of distance would cause the force of gravity to reduce by a factor of nine. This means that as the comet travels further from the Sun its speed decreases rapidly, with it having a very low speed at the furthest point from the Sun. As it comes back towards the Sun it will start accelerating again.

Pages 67–68

1. **Vector:** Quantity where direction is important, e.g. velocity, acceleration; **Scalar:** Quantity where direction is unimportant, e.g. mass, time.
2. (a) Relative speed = 6 + 8; Relative speed = 14 m/s. **(1 mark for calculation, 1 mark for correct answer.)**

(b) Relative speed = 8 − 6 = 2; Relative speed = 2 m/s. **(1 mark for calculation, 1 mark for correct answer.)**

3. (a) $v = u + at$; $v = 0 + (10 \times 3)$; $v = 30$ m/s. **(2 marks for calculation, 1 mark for correct answer.)**

(b) $s = \frac{(u + v)}{2} \times t$

$s = \frac{(0 + 30)}{2} \times 3$

$s = 45$ m **(1 mark for calculation, 1 mark for correct answer.)**

(c) $t = \frac{s}{\left(\frac{(u + v)}{2}\right)}$

$t = \frac{60}{\left(\frac{(5 + 35)}{2}\right)}$

$t = 3$ seconds

(1 mark for rearrangement and calculation, 1 mark for correct answer.)

4. $v^2 = u^2 + 2as$; $v^2 = 0 + (2 \times 4 \times 50)$; $v^2 = 400$; $v = 20$ m/s **(2 marks for calculation, 1 mark for correct answer.)**

5. At max height, $v = 0$ **(1 mark)**, $S = (v^2 - u^2) \div 2a$; $S = 0 - 20^2 \div (2 \times -10)$; $S = 400 \div 20$; $S = 20$ m **(2 marks for calculation, 1 mark for correct answer.)**

Pages 69–70

1. Trajectory
2. The range increases as the angle increases from 10° to 40°, but then decreases; Sarah's prediction incorrect; Maximum range of projectile when launch angle is 45°.
3. (a) $s = ut + \frac{1}{2}at^2$; $s = (0 \times 4) + \frac{1}{2}(10 \times 4^2)$; $s = 0 + (\frac{1}{2} \times 160)$; $s = 80$ m **(2 marks for calculation, 1 mark for correct answer.)**
   (b) $s = ut + \frac{1}{2}at^2$; $s = (7 \times 4) + \frac{1}{2}(0 \times 4^2)$; $s = 28$ m **(2 marks for calculation, 1 mark for correct answer.)**
4.

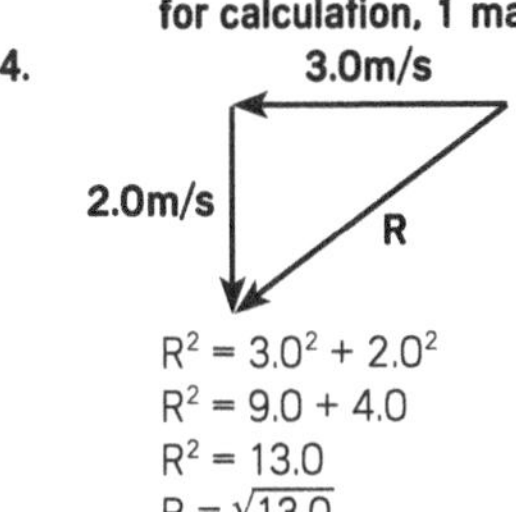

$R^2 = 3.0^2 + 2.0^2$

$R^2 = 9.0 + 4.0$

$R^2 = 13.0$

$R = \sqrt{13.0}$

$R = 3.6$ m/s **(2 marks for calculation, 1 mark for correct answer.)**

5. Horizontal velocity is constant (ignoring the effect of air resistance) as it's unaffected by gravity; Gravity causes the vertical velocity to change (increases).

Pages 71–72

1. (a) equal; opposite
   (b) Particles of gas in can; Gain kinetic energy from increase in temperature; This increases the pressure past the point the can walls can withstand and the can explodes.
   (c) Burning fuel creates gas molecules with very high kinetic energy; The particles collide with the rocket walls; This creates an upward force that causes the rocket to lift off.
2. (a) Recoil is caused by the conservation of momentum; Momentum before firing is zero; Total momentum afterwards must equal zero; Bullet carries momentum in forward direction so gun must have equal momentum in opposite direction.
   (b) As the temperature decreases particles have lower kinetic energy so travel slower; Impact with cylinder has lower change in momentum; Lower frequency of collisions.
3. (a) Total momentum before collision = total momentum after the collision.

$m_1u_1 + m_2u_2 = (1100 \times 10) + (800 \times (-10))$

$= 11000 + (-8000)$

$= 3000$ to the right

(2 marks for calculation, 1 mark for correct answer.)

(b) Momentum before = momentum after

$3000 = (m_1 + m_2)V$

$V = 3000 \div (m_1 + m_2)$

$V = 3000 \div (1100 + 800)$

$V = 3000 \div 1900$

$V = 1.6$ m/s to the right (accept 1.58 m/s)

(2 marks for calculation, 1 mark for correct answer.)

Pages 73–74

1. Microwaves from sat phone; Are transmitted to a satellite; Which then retransmits them back to Earth; Microwaves from ordinary phones don't use a satellite but require a nearby receiver.
2. (a) **Below 30 MHz:** Reflected by ionosphere.
   Above 30 GHz: Rain and dust reduce strength of signal due to absorption and scattering.
   Between 30 MHz and 30 GHz: Pass through atmosphere.
   (b) Between 30 MHz and 30 GHz; Best frequency to pass through atmosphere and reach satellite.
3. Receiving dish must be exactly aligned because satellite has a dish many times larger than microwave wavelength; Little diffraction so very narrow beam.
4. Sound has longer wavelength than light; Doorway is much wider than the wavelength of light; More diffraction when gap size smaller; Maximum diffraction when gap size is same as wavelength of wave.

Pages 75–76

1. electromagnetic; diffracted; interfere
2. (a) Reinforcement; Constructive
   (b) The light would be brighter.
3. (a) **Any two from:** Wave sources are coherent; Same frequency; Light is monochromatic.
   (b) Much brighter; Central fringe
4. Only horizontal oscillations get through; All other oscillations absorbed; Light dimmer (less glare).
5. Waves must have same amplitude; In phase; Same frequency.
6. (a) Constructive
   (b) Light twice as bright as single ray
7. As the polariser is turned; More light is absorbed by the second polariser so; Less light gets through; Once the second polariser is at 90° to the first no light will get through.

Pages 77–78

1. medium; speed; more; towards
2. **Any one of:** By how much it bends/changes direction; Change in speed.
3. $n = \frac{3.0 \times 10^8}{2.2 \times 10^8}$

$n = \frac{3.0}{2.2}$

$n = 1.36$

(1 mark for correct equation, 2 marks for calculation, 1 mark for correct answer.)

4. A
5. (a) Different wavelengths of light are slowed down by different amounts as they enter the prism; Higher refractive index, greater refraction.
   (b) Blue
6. n = speed of light in vacuum ÷ speed of light in medium; Speed of light in medium = speed of light in medium ÷ n; Speed of light in diamond = $3.0 \times 10^8 \div 2.4$; Speed of light in diamond = 1.25×10^8 m/s. **(2 marks for calculation, 1 mark for correct answer.)**

Pages 79–80

1. (a) Convex
   (b)

Focal point

Focal length

Principal axis

2. In cameras; Magnifying glass; Projectors; Spectacles.
3. Magnification = size of image ÷ size of object; M = 3 cm ÷ 1.2 cm; M = 2.5. **(1 mark for calculation, 1 mark for answer.)**
4. original height = image height ÷ magnification
   10 ÷ 60 = 0.17 m **(1 mark for substitution, 1 mark for correct calculation.)**
5. **Real**: Can be projected onto screen; Inverted.
   Virtual: Right way up; Cannot be projected on to screen.

P6 Electricity for Gadgets

Pages 81–82

1. **Any three from:** Speed of motor increases; Shorter length of wire between contacts; Less resistance; Greater current can flow.
2. **(a)** Doubles
   (b) voltage = current × resistance **(1 mark)**
   = 48 × 0.5 = 24 V **(1 mark)**
3. Component **A**; Graph is steepest (has greatest gradient)
4. **This is a model answer, which demonstrates QWC, and would therefore score the full 6 marks:** As current flows in wire electrons collide with atoms. This increases the kinetic energy of the lattice ions and causes an increase in vibration, which then leads to further collisions. This leads to an increase in the temperature of the wire and also its resistance. This increase in resistance prevents the current increasing further, which produces the shape of the graph shown. This is an example of a non-ohmic conductor.

Pages 83–84

1. **(a)** Total resistance, $R_T = R_1 + R_2 + R_3$; $R_T = 6 + 6 + 18$; $R_T = 30\ \Omega$. **(1 mark for calculation, 1 mark for correct answer.)**
   (b) Decrease
2. During the day high light intensity, low resistance; At night low light intensity, high resistance.
3. **(a) (i)** 6 V
   (ii) It decreases
   (iii) It will decrease
   (b) $V_{out} = \frac{R_1}{R_1 + R_2} \times V_{in}$
   $V_{out} = \frac{10}{10 + 20} \times 12$
   $V_{out} = \frac{10}{30} \times 12$
   $V_{out} = 4$ V
   (2 marks for calculation, 1 mark for correct answer.)
   (c) Control circuit to turn on lamp when light level drops, e.g. in a greenhouse.
4. When light is bright the resistance is low/so output voltage is low; When light level drops/resistance increases; Output voltage increases and buzzer sounds.

Pages 85–86

1. **(a)** An electronic switch.
   (b) High voltage (5 V); Low voltage (0 V).
2.

A	B	Q
0	0	0
0	1	1
1	0	1
1	1	1

3. **Benefits: Any two from:** Lower material costs; Higher production output; Potentially faster operation.
   Drawbacks: Increased complexity in circuit design leads to increased engineering and research and development costs; The components are not serviceable after manufacture (can't be fixed but must be replaced).

4. $I_e = I_b + I_c$
   $I_e = 0.1 + 1.3$
   $I_e = 1.4$ A
   (1 mark for calculation, 1 mark for correct answer.)
5. +6V
   A
   B
   Out
   (All correct for 2 marks.)
6.

A	B	Q
0	0	1
0	1	0
1	0	0
1	1	0

(All correct for 2 marks.)

Pages 87–88

1. **(a)** A, B, C, D, Q
   (b)

A	B	C	D	Q
0	0	0	0	0
0	0	1	0	0
0	1	0	0	0
0	1	1	0	0
1	0	0	0	0
1	0	1	0	0
1	1	0	1	0
1	1	1	1	1

(All correct for 4 marks.)

2. 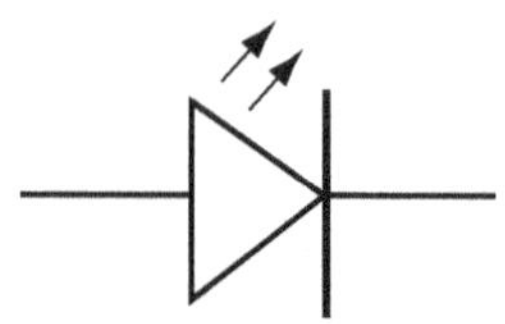

3. Small current in relay coil; Closes switch in high-current circuit.
4. Logic gate has low power output; Mains works at high power; Relay isolates low voltage from high voltage to avoid risk of shock.
5. **(a)** As light level decreases; Resistance of LDR increases; Output (input to logic gate) is high; Output of AND gate is high.
   (b) Swap the fixed resistor for a variable resistor.

Pages 89–90

1. **(a)** Same as field around a bar magnet.
   (b) Solenoid
   (c) Increase the current; Increase the number of coils.
   (d) Poles swap positions
2. **(a)** Magnetic field appears around the wire; This field interacts with magnetic field; Wire experiences a force, so moves.
   (b) Reverse the direction of the current; Reverse the direction of the magnetic field.
   (c) Place wire at right angles to magnetic field.
3. Force/wire pushed into page.
4. Motor uses direct current; Split-ring commutator enables current to flow; in one direction relative to the magnetic field.
5. Radial field; Coil always at right angles to magnetic field; Maximum force.

Page 91

1. Dynamo effect
2. 50 Hertz
3. **(a)** Bulb becomes brighter because voltage increases.
   (b) Bulb becomes dimmer because voltage decreases.
   (c) Bulb becomes brighter because voltage increases.
4. DC generator has commutators; AC generator has two slip rings and brushes.
5. **Any two from:** If coil spins faster, wires 'cut' through magnetic field lines more quickly; Wire experiences faster rate of change of magnetic field; Induced voltage greater; Frequency of output AC voltage increases.

Pages 92–93

1. heat; large; step-up; current
2. Step-down transformer decreases voltage; Has more turns on the primary coil than the secondary coil.
   Step-up transformer increases voltage; So has more turns on the secondary coil than on the primary coil.
3. Two coils not connected; User isolated from mains supply; Less risk of electrocution when near water.
4. **(a)** Step-down
   (b) $\frac{V_p}{V_s} = \frac{N_p}{N_s}$

   $V_s = \frac{N_s}{N_p} \times V_p$

   $V_s = \frac{50}{1000} \times 400$

   $V_s = 20\text{ V}$

   (2 marks for calculation, 1 mark for correct answer.)
5. **(a)** Power loss = (current^2) × resistance; Loss = $(4^2) \times 100$; Loss = 1600 W. **(2 marks for calculation, 1 mark for correct answer.)**
   (b) 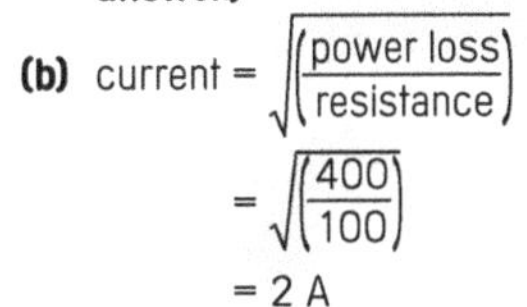

   current = $\sqrt{\left(\frac{\text{power loss}}{\text{resistance}}\right)}$

   $= \sqrt{\left(\frac{400}{100}\right)}$

   $= 2\text{ A}$

   (1 mark for calculation and rearrangement and 1 mark for correct answer.)

Pages 94–96

1. **(a)** (Silicon) diode
   (b)
   (c) Current only flows in one direction; As resistance in other direction is very high.
2. **(a)** **Half-Wave Rectified AC**

   \+
   Voltage
   Time
   –

   (b) A diode
3. **(a)** C
   (b) A bridge circuit produces full wave rectification.
4. **(a)** The capacitor would gain charge; The voltmeter would show an increase in voltage.
   (b) The current in the circuit will start high and decrease over time; As the capacitor discharges; Voltmeter shows a reduction in voltage over time.
5. Capacitor is used to supply a constant voltage; Capacitor is charged when input voltage is high; Capacitor discharges when the voltage is low.